Science of Rhythm

Indian System of Musical Rhythm "Taal Shastra"
Analysis of its Science and Sensibilities

I0468105

Bhavesh C. Bhagat

Author of Science of Melody

Foreward, Dr. Vishwambharnath Mishra

(Shri Mahantji Sankat Mochan and Tulsi Ghat, Benaras)

HoD Indian Institute of Technology, Benaras Hindu University, Chairman Swatch Ganga Foundation

First Edition, 2019

Publisher:

Universal Pilgrim Productions

Virginia, Varanasi, Vadodara

Color Paperback ISBN: 9781099787089

Monochrome Paperback ISBN: 9781072100607

Second Edition, 2020

Publisher:

Universal Pilgrim Productions

Virginia, Varanasi, Vadodara

pilgrimuniverse@gmail.com

Color Paperback ISBN: 9781099787089

Monochrome Paperback ISBN: 9781072100607

Lord Shri Krishna with Shri Radhe....Divine Rasa Lila....

Rhythm of Mridang (Taal vadya) and Jhaanjh (cymbals) with Sakhis

Forests of Vrindavan, Brij, India

DEDICATION

I bow humbly to Lord Shri Krishna and his eternal Lilas full of "Rasa" encompassing all Universal Music in His "Naad Brahma".

This small volume is laid out in English for the first time and discussing the intricacies of science and sensibilities of one of the oldest form of Rhythmic Structures (Indian Taal Shastra). It is with deepest love and gratitude dedicated to Lord Shri Krishna and all the musically inclined souls in the universe, persevering in their humble effort to resonate their frequencies in "HIS" "Swara" and "Laya".

Dedication of this effort to Lord Shri Krishna would not be possible without the grace of my spiritual Guru, Vallabh Vedantacharya Shri Shyam Manohar Goswamyji (Shyamubava) of Kishangarh and Mumbai, who has inspired me to sing with him often in Kishangarh and to proceed and progress expediently on the path of seeking and practicing all aspects of Music in service to our Lord's Lotus feet.

रसो वै सहः

(The nectar of Infinite Joy of Universe is the Lord himself)

"Venu Naad" in Khinmey Monastery, Tawang, Arunachal
Pradesh
Photo by Author

Contents

FOREWARD SHRI MAHANTJI, SANKAT MOCHAN 10

PREFACE **18**

ABOUT THIS BOOK **28**

BHAVANJALI _ACKNOWLEDGEMENTS **32**

VALLABH VEDANTACHARYA SHRI SHYAM MANOHAR GOSWAMYJI33

NL SHRI MUKUNDRAY GOSWAMYJI, MUMBAI 36

SHRI MAHANTJI VISHWAMBHARNATH MISHRA.....................38

MAHANTJI SW. SHRI AMARNATHJI MISHRA (BABAJI) 41

PADMASHRI DR. RAJESHWAR ACHARYA "PRABHAVRANG" 45

SW. SHRI MANNUJI MRIDANGACHARYA 47

SW. MRIDANGACHARYA SHRI RAMSHANKAR PAGALDAS 48

NL SHRI DEVAKINANDAN GOSWAMYJI INDORE 50

'RASIK PRITAM' SHRI HARIRAY MAHAPRABHU 51

SCIENCE OF MUSIC - RHYTHM AND NAAD BRAHMAN .**52**

SCIENCE ART PHILOSOPHY RELIGION 52

MUSIC - PHILOSOPHICAL EVOLUTION 55

ANAHAT NAAD 55

AAHAT NAAD 56

JOY IN MUSIC – PHILOSOPHICAL AND SCIENTIFIC "RESONANCE"..57

NAAD PSYCHOSOMATIC PRODUCTION. 59

SCIENCE OF RHYTHM...62

LOK RANJAK AND ATMA RANJAK AAHAT NAAD 63

NAAD YOGA AND TAAL SHASTRA 66

MINDFUL MEDITATION IN MUSIC 68

REVERSE ENGINEERING MUSICAL PROCESS 69

NEUROSCIENCE AND RHYTHM 73

SECTION 2 - TEN PRANA OF TAAL........................**77**

THE PRANA TABLE..78

TAAL AND ITS NATURE...81

KAAL, MAARG, AND KRIYA 87

   Kaal..*87*

   Maarg..*90*

   Kriya...*91*

ANG, GRAHA, JATI...94

   Ang *94*

   Graha (Sama, Atit, Anagat)...*102*

   Jati...*106*

KALA, LAYA, YATI, PRASTAR 109

   Kala...*109*

   Laya (Vilambit, Madhyam, Drut) *113*

   Yati..*118*

   Prastar...*131*

SECTION 3 - PRACTICAL CONCEPTS............................**135**

TIHAI, MOHRA, BOL, PARAN, TUKDA 135

   Tihai..*135*

   Mohra..*137*

   Bol..*138*

   Paran...*138*

   Tukda *139*

PALLU, CHAUPALLI, ANGUSHTANA, FARAD 141

   Pallu..*141*

   Chaupalli...*141*

   Angushtana...*142*

   Farad or Ekkad...*143*

GAT, BAANT, PENCH, RELA, LADI, LAGGI, UTHAAN....144

   Gat...*144*

   Baant...*144*

   Pench *144*

   Rela *145*

   Ladi (Dhanakshari Chand)..*146*

   Laggi *146*

   Uthaan (Amad)...*147*

SECTION 4 - ADVANCED CHAND FORMATIONS 148

CHAND BHUJANG PRAYAT..149
JHULNA CHAND (THREE VARIETIES) 149
"ASHWA GATI" CHAND..151
SHIKHARINI CHAND..151
MANHAR CHAND 151
DHANAKSHARI CHAND 152
DEV DHANAKSHARI CHAND...152
DANDAK CHAND..153
TEVRA CHAND IN DOUBLE SPEED......................................153
GANDAKA CHAND 153
CHAMPAKMALA (KEHERWA) CHAND 154
CHARCHARI (JHAPTALA) CHAND 154

BIBLIOGRAPHY 156

ABOUT THE AUTHOR...157

OTHER BOOKS BY AUTHOR 159

ABOUT THE PUBLISHER..161

FOREWARD Shri Mahantji, Sankat Mochan

(Dr. Vishwambharnath Mishra)

Rhythm – Its very presence is as essential to our everyday lives as one of the key ingredients of life force - our breath. One could be breathing but without rhythm, that living organism is essentially devolving. Take the human heart as an example, which is the crucial muscle operating inside our bodies, which derives its energy from oxygen in our breath and performs the essential task of pumping blood and life force in our veins. This act of pumping itself must occur in specific Rhythmic bands. If there is even slightest irregularity in this Rhythm of heart, our physical well-being is immediately affected. Therefore, the living organisms have this inbuilt urge and need provided by nature to "Maintain" their inner rhythm. Human beings try all sorts of ways via modern techniques and exercise to maintain their Heart's Rhythm for this purpose. The resting adult heartbeat range is between 60 to100 and often it is ideally around 72 BPM (Beats per minute). It is not an accident that in the Science of our Indian Rhythmic Structures in Mridang (the oldest actively used Indian Rhythmic Instrument) the main base rhythmic structure is of 12 beats (Chautaal) which is a multiple of 72 and leads to our resting Heart BPM rate.

These laws of natural rhythm do not just apply to our bodies but also in all aspects of our lives. For example your work, your singing, your exercise, etc. Therefore, whatever process you are doing in any activity in life such as Musical singing for example, if this natural rhythmic balance in that act is not maintained, that song will have no impact and music will be less than ideal. This fact is lucidly highlighted by our ancient Acharyas and Sages, as Shri Tulsidasji describes in his Shri Ramcharitmanas.

रामनाम जपहिं लय लाए। होहिं सिद्ध धनमादिक पाए।

There is a direct reference to "Laya" and Rhythm here by Shri Tulsidasji who says that even singing (Jap) the Lord's name when it is done in "Laya" then the person singing achieves many "Siddhi's" (extraordinary feats) without making any extra effort. This is the power of Rhythmic balance. So if you sing anything be it even Lord's name sing it in Rhythm. What occurs when this "Jap" or "Kirtan" is done in rhythm is that one naturally then finds a means of connecting with the "Power Center" of the Creator or Providence. Thus taking Lord's name even without Rhythm actually means nothing it must have Rhythm.

Spiritual Rhythm - Thus Rhythm, as we can see, is essential to our daily lives. In a common act of a person singing, it goes without saying that all our Music is built on the foundation of Rhythm. In our ancient Indian Classical traditions, Music is not just an act of entertainment for humans. Yet when we sing it gives us joy and often we feel happy but that is one of the side effects of the act. Our thousands of years old culture has always treated music as a means of Divine Connection. The importance of Rhythmic structures is also evident from our spiritual practices continuing to this date. In the Indian system of worship when one interacts with a Deity, there is by and large a system of "Raag, Bhog, and Shringar". This occurs in many different forms but at a macro level in our temples where the Deity is worshipped they are offered Food offerings (Bhog) that devotees then partake as prasad. They are offered adornments (Shringar) and they are offered Music (Raag) in form of many specialized schools of Spiritual sub-systems of Music based on specific formats of Rhythms. This is the Temple format. However, there is one very important another format of Worship that is even practiced.

In the traditions of a Householder such as the Vaishnavite traditions, the emphasis is on considering the Divinity as part of our own family. The Deity is treated as one's own family member (in the role of a friend, child or any other relation which suits the devotee's sensibilities) Here the Raag Bhog and Srinagar take on a new meaning because they are personalized directly between the Deity and the Devotee. There are no intermediaries here and the emphasis is on offering the best to the Lord. Therefore, the daily life of the family is shared with the Deity in exactly the same manner as we live but with an emphasis on offering best and purest ingredients to the Lord. In this unique Grihasth (Householder) Parampara, just like our routine of daily life, one wakes up the Lord, then one offers him or her the breakfast (Bhog) after bathing and getting ready (Srinagar). Then the Deity is offered Lunch (Bhog) around noon and that Lunch is then partaken by devotees' family as Prasad. Then Lord takes a nap in the afternoon and the evening cycle of daily life takes on until Lord sleeps at night.

While these specific interactions are occurring between the Deity and the Devotee, the Music is also sung based on the occasion (Raag) such as in Morning prior to waking up the morning Ragas are sung in Vaishnavite systems. Then at Lunch, afternoon Ragas are sung for the Lord with Mridang or some other Rhythmic Instrument. At wake up time from a nap in the afternoon, soft afternoon Ragas are played on Veena and so on till at night when the Deity is going to sleep (Shayan) the night Ragas such as Bihag are sung to soothe the spirits.

This whole Daily Cycle of devotion "Nitya Krama" process has a structure and its own Rhythm. You cannot just randomly wake up Deity at midnight and offer breakfast just as you yourself would not eat breakfast at midnight. The daily tempo and Rhythm of life must be maintained in this spiritual interaction. The "Laya" or balance in this format is very important. So if your try to interact with your Deity and this interaction and your process of practice is out of Laya and not consistent and persistent, then instead of getting closer to your Deity, you are putting more imbalance and distance between Devotee and Deity. Thus, Rhythm has paramount importance in everything we do materially and spiritually.

In addition, it would be remiss if we do not remember Shri Hanumant (Hanumanji) now. Lord Hanuman himself is an Eternal Devotee Vaishnava. It is said that there is no greater devotee in this living universe than Shri Hanuman. He is the greatest devotee of Lord Shri Rama and it is said about him in our scriptures that;

"अतुलित बलधामं हेमशैलाभदेहं दनुजवनकृशानुं ज्ञानिनामग्रगण्यं"

He is the Leader of all the Knowledge amongst all the Learned. He is the Strongest amongst all living beings. So these two traits define whenever we discuss anything about Shri Hanumant. One is the Knowledge and the other is Strength. He applies both these qualities to become a perfect Devotee of the almighty Lord Shri Rama. He is also the founding Acharya of one of our root Musical schools of thought amongst other sages. His Musical philosophy is called the "Hanumant Mat". He is so much devoted to singing his Lord's name that it is said his every being and every atom is profused with the musical character of Lord Shri Rama's Naam Kirtan. It is due to this strength and his commitment that Lord Hanuman is blessed to be with Lord Shri Rama in every future manifestation until eternity to perform his devoted service.

Lord Hanuman with his own actions and his own Music inspires to perfect our own devotions to the Almighty. Our bodies are impure and thus our pursuit of Music is often incomplete and often full of mistakes. Nevertheless, if we

13

have true devotion to offer this Music to our Lord, then Shri Hanumanji himself becomes the Guru, he is the eternal Devotee and he guides forever the souls wishing to sing the Lord's name in perfect Rhythm and Melody.

Musical Rhythm– In our Music what do the above explanations mean? It means that we must strive to always retain our balance of Rhythm "Laya" and we must retain our harmony of Melody "Swara". This synchronization is easier said than done in real life. It comes with many years of conscientious effort and persistent practice. Even one life is not enough as one tries to perfect the Swar and Laya and continue to offer the best ingredients of Music to our Lord. The moment you achieve this perfect synchronization of Swar and Laya in that union itself one would realize the glimpse of their Lord. Thus the beauty of both these ideas is that if one is singing in "Swar" and a percussionist is creating a rhythmic base then the Percussionist also must be in "Swar" in his mind, body and his instrument of percussion. On the other hand, the singer who is singing must be in "Laya" and balance with the Rhythmic structures being presented to them as per the pre-agreed format of the design of that composition. Thus, this harmonization has to be constantly achieved between various performers.

Since the present Author and I both are, Electronics Engineers let me reference here an Engineering analogy. We have these communications channels in engineering. These channels are each defined within a specific bandwidth (range of frequencies). If the signal is transmitted for communication falls outside the range of a given defined channel and its band of frequencies one can try as hard as they can but it would never transmit on that channel and hence never communicate. Therefore, what occurs in Music is that the band of frequencies between "Laya" and "Swara" must be observed and performers must communicate in the given predefined channels. These are the channels in which we perform our music but one must never forget that there is an overarching channel of band of frequencies where the Divinity performs their music and our ultimate goal is to resonate with that channel and communicate with that through the path of our Music.

What is essential in good Music is Melody. However, Melody alone is not sufficient. The Melody must come continuously and must not stop, it must keep flowing. Consider the mechanism of this flow of melody. The Waveforms in which the melody comes and repeats itself in specific cycles make different combinations and melody unique. These cyclical orders of repetitions are nothing but the second most essential quality of Music and that itself is Rhythm. Then what is the "Sur" or "Swar" in this context? Swar defines the actual body or portions of Melody and these portions are repeating in a cyclical order of Rhythmic waveforms. Therefore, the frequency of Sur and Rhythmic Waveforms must match to produce resonance, which we identify as good Melodious Music.

A Singer when he or she sings a given Raga it is nothing but these permutations and combinations of these modifications of their speech according to the frequency of the predefined structural mechanism of "Swar" of that Raga. Inside this structure, the Singer has full variability to create continuous Melody however how will that melody reach the Listener. There must be a communications channel and that channel is defined by the Rhythmic Cyclical Order. The Bandwidth through which melody would be appreciated in that context would be established by the Rhythmic accompanist who would repeat and create the Cyclical Order of Rhythm in which Melody can travel and continuously expand. Now if the Rhythm player cannot balance and maintain the frequency of his Rhythmic Cycle (his/her Laya) or vice versa the Singer can't sing to be in bounds of the waveforms of the rhythm then in both cases we would have miscommunication of frequencies causing disharmony and disruptive melody. The compatibility between two, therefore, is essential to producing good "Sangat" and good Music.

Summary

Summary –All of this Musical Process is a spiritual Feedback loop mechanism. When one wishes to strive to perform and offer music to our Lord the question is with what mindset are you offering this music to the Lord. You might be the best musician but if your mindset is clouded by public fame and material thoughts when performing music in front of the Lord it is one thing; perhaps not suited for the offering. However, if your mindset is pure and full of humility and even if you are not perfect as a Musician, then the Lord will listen and guide. We must realize when we strive to offer our Music to the Lord that he is the Perfect Creator of all music. He is the all-knowledgeable creator of all Arts. So when we perform with Humility for him, he will identify mistakes and start self-correcting for you and send you the right guidance needed as he only creates and listens to the perfect music. He will ensure that the right Guru comes and holds your hands at the right moment to continue this eternal journey of "NaadBrahma"

So ultimately what is the purpose of all this? **It is to transform our Music from an Art to _Ingredient_.** An indispensable and irreplaceable loving ingredient to offer to our Lord and in that process, it will lead to infinite Joy. The whole process takes us to the realization that "since I am his integral part but have somehow forgotten that knowledge, I should again become his integral part". In our Kashi (The City of Liberation or Moksha), every soul comes with a desire to reconnect with that Infinite Joy our Creator. Our goal then is to use humbly our persistence and humility led efforts on the path of Music to connect us back to the Supreme. Those who understand this truth, the world will automatically come to listen to those souls because they are slowly but surely able to maintain the balance of their Rhythm and Melody with the Supreme in the daily acts of offering the Music to the Lord. Their infinite Joy is in offering the Music at feet of their Lord not in performing for the World. Then a stage comes when the performer can see the Creator in all his Audience and everything becomes an offering.

The Rhythmic traditions that I represent are steeped in Spiritual and Philosophical foundations as described above. The seeds of our Benaras Tulsi Ghat spiritual tradition of Mridang and its Rhythmic practice were sown by my Grandfather Shri Babaji Mahantji Amarnathji. This tradition is now being continued by our humble efforts under the guidance of Lord Hanumanji. I am very happy that our tradition of Mridang Vadan is alive and thriving in the present efforts of souls like Shri Bhavesh Bhagat from Virginia (USA) and Vadodara (India). It is said that distance and separation increases the passion and intensity of love. Bhavesh has lived more than half his life in the USA and it is only with the blessing of the Lord that his passion for our Indian Philosophy and Music continues to intensify even further considering the challenges of living in a modern fast-paced western world. Observations put forth in this book by the Author are directly arising from his own unique perspective seeing and experiencing the Indian Music and its Philosophy from outside as a neutral observer and as an experimenting practitioner and a lifetime student. This effort to present in the English language some of the essential meanings of our Science of Indian Rhythmic Cycles would prove to be an indispensable reference for any serious practitioner "Saadhak" of Indian Music. I wish the author and all those souls who open these pages continued blessings from the Lord to succeed in their individual musical journeys and may they all find their "Laya" in their "Naad Yatra".

Vishwambharnath Mishra

Kartika Chaturthi 2019

"Naag Nathaiya" Festival Benaras

PREFACE

Dear Friend,

 If you have come this far after picking up this book and initiating to read the Preface, one can safely assume that you have a heart that desires to learn more about Music and Rhythm. *There are many reasons to write this volume. Most importantly, if even one soul gains some inspiration and knowledge, using the words written here as a reference and applies it in the practice of Indian Rhythmic Science and Art towards a spiritual goal, then the author's effort would have been worthwhile. "Saarthak"*

There is also a "personal joyous gain" in this effort of writing. One's learning and knowledge is not solidified until one puts the ideas in writing and reinforces them in one's mind and soul. This has been one of the reasons why the pursuit of writing this book has benefitted this author in his Taal studies. This effort and the process of writing this book has assisted in further reinforcing and refining the scientific practice of musical knowledge in the author's mind. Besides this, in any culture, those who do not share the traditional ancient Knowledge that they have been graced with, ultimately do a disservice in their own progress.

The present author has been listening to the various forms of Indian Classical music for many decades and has had the fortune of not only persisting in this effort but also to graduate from being a passive listener to being a continuing lifelong practicing and experimenting student of Indian Classical music in its devotional form. The challenges along this path have been numerous but the efforts have been worthwhile for the soul's progress. Having been born in Mumbai, India, the author has spent over half of his life now in Northern Virginia near Washington DC in the USA. This distance from the root of the Indian Classical Music has given a renewed vigor and persistence in the author's Musical journey with all the challenges that geographical separation brings. However, this distance also is a blessing in that it keeps the mind away from all local distractions. This helps the author tremendously in staying focused on the true purpose....to practice and excel in all aspects of Indian Classical music and to offer it as an irreplaceable ingredient in the Author's devotional process.

The present state of Indian Classical Music is not in the original form that the founding Musical Sages envisioned it to be. However, just like life itself, Indian Classical Music has evolved and its purpose and its practitioners have evolved. This is good in a way because if art does not evolve it slowly withers away. Nowadays a sincere thirst to go to the "root" of all music is often missing in the practitioners and listeners both. In our case, it is the root of the Taal and its structures. What we find in Indian Taal knowledge now is rote knowledge passed on by modern *"Expert" (Pandit)* musicians focused mostly on the outer world. The "Spiritual" practitioner who takes the challenge to focus on the inner world, certainly also exists but is hard to come by, and for folks like the author, living far from Indian shores, this is even rarer. The performances given on stage in big cities offer music, but this is often superficial and more about a show of physical prowess for the most part. The performer is trying in the back of their mind, to think of how to obtain another performance, how to get more people to clap, or how he/she might look on social

19

media. Some of this is necessitated by our modern society and is essential even for survival; however, that "thirst" for understating the "roots" is missing.

The Student-Teacher, that is "Guru – Shishya" parampara of passing down traditional cultural assets such as Music, also has been hampered with vagaries of modern society, ignorance and often materialistic mindset of the teacher as well as the student. Often, the teacher does not teach the full depth of the subject leaving the inner teachings for his progeny (son or daughter) to keep their material source of income going. And many of today's students also are driven by the urge in our world of instant social media selfies, to become *"famous first and experts later"*, ignoring the patience and sacrifices required to hone a passionate art such as Indian Rhythm or any form of classical music.

Often the commitment and persistence in the Guru and Shishya are lacking today. Mostly, because the end goal is not the "Spirit", it is not the "Devotion" but it is rather "Ego". This is beyond our control; this is the reflection of our state of society and its evolution. Being "professional" in something as important as Music is a very commendable and most desirable thing in life to achieve. ***However if the "professional" is at the expense of the "spiritual" and "emotional" aspects of Music and its educational journey, then something is misplaced in that experience.*** Music taught or performed with the right "spirit" and purpose can be felt by the "attuned" soul. This is rare. Such spiritually inclined "Naad Yoga" teachers do exist in today's time and we wish them all the strength from Lord in continuing to pass on their spiritual music to their deserved students.

Indian Music has two components, "Swara" and "Laya". The business of Swara is conducted in non-rhythmic instruments e.g. "Vina Vadan" and vocal "Gaan", while the practical execution of Laya is conducted through rhythmic instruments and "Taal". Both these aspects of Indian Music have lost their original spiritual essence in modern mass performances with the emphasis being put mostly on the physical excellence of the player. The audience, for the most part, does not want to invest their time to learn about the nuances and depths of Swara and Taal. Such an audience "entertains the performer's ego" with fleeting claps and "Wah-wahs" and the show goes on.

What we find in the present modern musical performance, is that the "Taal" aspect, which is the keeper of the Laya, is almost treated as a stepchild on stage when in fact (as you will see in our book), "Taal' is the foundation for all music. The material considerations and incentives of the modern musical performances for famous musicians emphasize mostly the Vocal or Instrumental aspect of music treating Taal only as a secondary aspect. The listener too is now trained to think of Taal as something demonstrated on a side instrument. (Another instrument that has also suffered this similar fate like Taal is the accompanying "Lehra" instrument of Sarangi). Taal thus has seen more of its historical origins being lost to the vagaries of modern times.

Another technical reason for this is that Taal and its design by definition is the foundation on which the entire Vocal, Instrumental Music and Dance are built. Just like when a building is constructed on a foundation, it is often easier only to see the structure on top and completely miss the importance of the solid foundation underneath. To appreciate Taal in its singularity often the audience and practitioner do not have the depth to have a conversation in the performance as to what the Taal "Varnas" Spoken words of Taal sounds and its inner meanings represent. A vocalist can sing through the common language that audience can understand in song, an instrumentalist can demonstrate the melody. *Indian Rhythm demands a slightly higher standard of awareness, curiosity,*

21

and effort to understand its tricky words and language that is different from the perception of a common listener. So beyond tapping their feet to a natural human appreciation of a good beat as a base understating, most audience and practitioners do not venture further deeper into the exploration of the subject of Indian Classical Rhythm. These are a few of the observations about why Taal has become a secondary aspect of modern Indian Classical performance.

The question for the true seeker then is to find out how to progress in their Musical journey in these challenging but constantly evolving modern times. The big challenge is how to appreciate the original meanings of the minutest details and "Bhava" emotions of nature that our ancient Sages documented when they created the Rhythmic structures of Taal and Swara compositions of Raag. This is why the present author is embarking on this effort to illustrate in English the Taal Science and its intricacies to provide a meaningful reference to a student who is already playing some form of rhythmic Indian instrument such as "Mridang" or "Tabla" or for an "Active" listener who wants to make genuine efforts to grow their understanding of what is being performed in percussion and what their ears are hearing and their minds are comprehending.

This author has been fortunate as he grew up in India and honed his interest in reading old texts and over the years attained the ability to read in Hindi, Sanskrit, Brij Bhasha, Gujarati and Marathi languages. This, coupled with the blessings of many spiritually and musically oriented Gurus, has allowed the author to collect a library of some out of print and rare manuscripts on the arts of Indian Music. While some of today's modern music teachers often incorrectly guide the students by telling them to ignore writing the instructions and only to listen and play, etc., these are just half-baked truths that could hamper the progress of a true seeker.

22

Of course, memorization of "mind and the muscle" is the fundamental aspect of the Indian Classical musical system. As stated earlier, writing down instructions, reading, and learning are equally as important as listening and practicing. Writing and sharing one's written knowledge are one of the key ingredients of retaining and expanding one's knowledge of any subject. Each soul is unique in its own learning, evolution, and pursuit of this science and art of Music. One must find the right mix of guidance to persevere and keep progressing if one is committed.

Author's ongoing journey starting as an avid listener of "Swara" Melody and Rhythm "Taal"...and playing Mridang... to writing this book ...is connected with Benaras and Temples of Brijbhoomi and Rajasthan, India. Benaras (Varanasi) is the author's Musical home, while Kishangarh (Rajasthan) is his Spiritual home. The author had no relation to Benaras except for reading about it in texts, about it being one of the oldest cities in the world and the center of Indian philosophical learning since the early millennia and founding of the Indian cultural origins. It was almost two decades ago, in the early 2000s, that the author learned accidentally that on the banks of holy river Ganges, is hosted an International Dhrupad Festival on Tulsi Ghat. Mridang (Pakhawaj) which is the author's chosen instrument of his soul for Taal studies is played in Dhrupad and so he ended up in Benaras in 2005 for the first time pulled by the sound of the Mridang from shores of USA. This International Dhrupad Festival is hosted by author's Mridang Vidya Guru Shri Mahantji Vishwambharnath Mishra and his Sankat Mochan Temple foundation in collaboration with the King of Benaras foundation since 1975.

Sunrise. Tulsi Ghat. Ma Ganga - Photo by Author

For over a decade now, this author has been regularly visiting this festival to listen, learn and continue his hunger to play the Mridang and learn more about the science of Taal. **This body of the author had no foresight to choose Benaras as his destination, it is actually the city and river Ganges** *(on whose blessed Ghats the whole ethos of Indian life can be seen played out daily)* **who chose this author to pursue his Laya in the holy environs.** It is in this wonderful multiple day (all night long) festival that this author gets to observe every year along with challenges and opportunities faced by modern contemporary Indian Classical musicians. One gets to observe, listen and study amongst connoisseurs and practitioners congregated at Tulsi Ghat, Benaras from all over the world; Russia, France, Japan, S.Korea, Nepal, Bangladesh, UK, US, and many more. For such a seeker who travels this far to listen and learn, these ancient texts about Swara and Taal that have been published in the past can act as an entry point of curiosity to learn more.

This book is for the souls who gravitate from all over the world, attracted to Indian music. If this small reference volume gives such worthy seekers some essential flavor on the origins of Taal science and its format, author's effort would be worthwhile. Music is indeed a universal language.

The authors of the old rare books on Taal and its Science and practice were not just musicians and Taal experts, but above all devotionally inclined souls who undertook their writing efforts for future generations and in service of their God. In practical terms of playing a percussion instrument and understanding its language, the present author has benefited immensely in his progress as a student of "Taal Shastra" from two such souls who made such efforts and have guided us with their pioneering (now out of print) publications on Taal and Mridang. They decided to write for souls like this author to find their teachings and inspire them to continue their loving efforts.

The first is **Sw. Mridangacharya Shri Ramshankar Pagaldas (Pagalbaba) of Ayodhya**, who wrote with love from the depths of his heart, the *"Mridang Ank"* from Sangeet Karyalaya, Hathras, which has been author's textbook through which he kept his Mridang thirst alive over the years.

The second such soul is **Sw. Shri Mannuji Mridangacharya of Gopal Mandir, Benaras** who's rare and out of print *"Taal*

25

Deepika" series (published in 1930s) this author was fortunate to obtain. The copies were literally rotting under a mattress from a house in Varanasi years ago in early 2005.

These two souls are examples of the original and rare devotionally inclined teachers of Indian music tradition. They made efforts to write down these teachings so that future generations can benefit. It is the author's fortune that both these souls have offered their art and their science to Lord Shri Rama and Lord Shri Krishna respectively. Both Mannuji and Pagal Baba's souls would be pleased that every time this author's body plays, the "Thapiya" and its Laya are offered at the feet of the same Lord Shri Krishna.

There is a lack of authentic English literature on this subject and the modern student is more inclined to learn first by reading in English then proceed in other avenues, as such, this present effort to publish in English is essential. This author is joyously continuing the above tradition of sharing Mannuji and Pagal Baba's devotion by translating some of the knowledge obtained from Gurus and experiences from his own experiments and practice in English for future Western and Indian audience. This effort is imperfect, as Hindi and Sanskrit cannot be translated accurately, especially in English, but it is a beginning. All possible imperfections are authors alone and not of the original Gurus who shared their knowledge in Sanskrit and in Hindi.

The initial Scientific and Spiritual connections of Taal and Naad Brahma are written by the Author from his continuously evolving practice and studies of philosophy and music. All practical and advanced Taal sections representing Sanskrit and Hindi pieces of the present volume are compiled from aspects of the "Taal Deepika" four-volume series of Shri Mannuji Mridangacharya. Mannuji also happens to be the initiating point of Mridang Vadan, for the founder of the author's Mridang Guru Parampara of Sankat Mochan - Mahantji

(Babaji) Sw. Shri Amarnathji Mishra of Tulsi Ghat. This is the starting point and if Lord wishes, more such invaluable material would be translated and authored in English soon for future generations. More details about Gurus who have blessed the author with their kindness and knowledge and inspired him to pursue this pilgrimage of Naad (Swara and Laya) are acknowledged in the "Bhavanjali" Acknowledgement section.

Temple Verandah in Kishangarh, Rajasthan

About this Book

The "Why" and "What" of this book has been covered somewhat in the preface. Now let us take the opportunity to describe "Who" this book is for and "How" it is structured.

"Who" will benefit from this book? - The present volume is an English synopsis with some of the Author's own understandings as received from his Gurus over the years. The main source of the Sanskrit and Hindi material is out of print and unavailable "Taal Deepika" published by the Author Shri Mannuji Mridangacharya of Benaras as described earlier. Shri Mannuji was a scholar and has summarized in his life's learnings, all Taal texts from olden times into this volume. The original Sanskrit and Hindi explanations have also been maintained in this book to assist the reader in grasping the idea from multiple languages.

The book deals with the subject of illustrating the Science (logical and experimental roots) and Sensibilities (emotional and spiritual feeling based genesis) of the Indian Rhythmic structures also known as "Taal Shastra" in Sanskrit. Like all forms of Science, some preexisting background and conditions have to exist for a student to understand the postulated theories and the experiments. The Science of Rhythm is no different.

1. One must be curious to explore and be willing to understand the depths of the Rhythm and its structures beyond the physical aspect of playing the instrument.

2. One must be familiar with some basics of playing a percussion instrument such as Mridang (Pakhawaj) or Tabla. This book is NOT a guide on "How to Play" the instruments. That subject would be covered in future volumes and is to be experienced by practically playing

the instrument preferably under the guidance of a Guru well versed in Taal Shastra.

3. If no such Guru is found, the seeker is to continue self-progressing and self-learning the physical aspects of the instrument by "Active" listening and practicing without fear. By making an effort to find and learn from some of the written materials as this Author has found. When the right time comes, Guru will present himself or herself, what is essential is the commitment, and there can be no better proof of truth in this eternal secret principle than the life of the present Author itself. A true seeker of Music is fearless when they are embarking on the journey as an "offering to the Lord."

4. The reader must have a spiritual and emotional bent to appreciate the dimensions of music that cannot be described in a book but rather can only be experienced in solo meditative practice.

5. The Reader must be willing to experiment with the ideas. Understand even some basic points and connect them with his/her own personal practical Rhythmic experience and seek the guidance of a Guru who will assist in continuing this wonderful exploration further.

6. While the knowledge of Sanskrit and Hindi is not essential for 80% of the contents, some basic curiosity about Hindi words and their meanings would only further enhance the Reader's understanding and is especially needed for the remaining 20% Advanced sections of the book in later part.

"How" is the book structured – The main ideas are covered in four sections.

1. The first section deals with the actual analysis and spiritual origins of the concept of Music and Naad Brahma. This is followed by the psychosomatic process of the creation of music and rhythm.

2. Second section then dives deeper into the Ten Life Force (Prana) of Taal. This lays the foundation to explore Indian Rhythmic science where the Ten key characteristics of Taal are illustrated.

3. The third Practical terms section deals with some modern-day ideas evolved from the original science of Taal and the terms used to define the practical structures in use today. Familiarity with both of these would further enhance a seeker's depth in the exploration of the Taal Science and its Art.

4. The Fourth and final Advanced section about Chand and the pieces where Taal examples are given are in Hindi, as translating them in English would be counter-productive to our macro effort. These sections are for the student who is already a little familiar with Hindi and can seek guidance from a learned master in applying the concepts in practical experimentation and creation of the new ideas.

This is a body of work that is continuously evolving as the author and his journey of Naad Yatra "Pilgrimage" evolve. As such, the second and third sections of the book on the Taal Shastra from old ancient texts will remain firm in their definitions and will remain as a steady reference for the future. However, the application of scientific principles and emotional ("Bhavatmak") sensibilities of practical use of Taal Shastra from a spiritual perspective will continue to evolve. As the author enhances further understandings from his experimentation and from Gurus' blessings, future editions will be revised accordingly. Hope is that this small volume will be of useful reference for practitioners and listeners alike;

those souls worldwide who are thirsty for a deeper understanding of Indian Classical systems of Music.

Bhavanjali _Acknowledgements

The contents and every word and feeling in these pages are not this author's but rather inspired and acknowledged as the blessings of great souls mentioned in this "Bhavanjali" section. *As such this author list his Spiritual Guru Shri Shyam Manohar Goswamyji and Mridang Vidya Guru Shri Mahantji Vishwambharnath Mishra as his co-authors along with Shri Mannuji Mridangacharya who originally made the invaluable loving effort for sharing his "Taal Deepika" with future generations.* What is being written and being published is what the author has gained from his blessed Gurus and is sharing with all.

Author wishes to thank the blessings of his Mother and Father along with other elders who have planted the seeds of this joyous journey in his life. Also it would be remiss if the author did not thank the gracious souls who have helped with their various efforts to proofread the manuscript and assisted in some way or other in creating a volume with hopefully minimal errors. Author's daughter Pushti Bhagat who is a student of Indian Bansuri and son Brij Bhagat who is a passionate student of western classical Violin; for encouraging and assisting in these efforts. Mr. Sushant Karmarkar and his father Shishirji both for their help in creating a custom instrument for the author. Both are also accomplished Dhrupad singers in their own right. Pandit Badriji and "Dada" Arun Chatterjee of "Gurukul" music school at TulsiGhat who always are eager to spread the joy of Laya with Shri Mahantji. I wish that the Lord grants them their continued success in the pilgrimage of Music.

All errors, omissions and shortcomings are author's alone in writing and compiling this work. The mind and our bodies are full of imperfections. This pilgrim asks for grace of the great gurus illustrated here for their continued blessings onwards on this "Naad Yatra".

Author's Spiritual Guru

Vallabh Vedantacharya Shri Shyam Manohar Goswamyji

Shri Shyam Manohar Goswamy (Shyamubava) of Kishangarh, Rajasthan and Vile Parle, Mumbai

Author receiving blessings of his Diksha Guru

Shri Shyamubava represents the 16th Generation in the lineage of Mahaprabhu Shri Vallabhacharya (the founder of Shudh Advaita Vedant philosophical school of Hinduism). He is the foremost scholar and a great literary author in the world in the philosophy of Pure Non-Dualism with worship of Lord Shri Krishna at its heart. He has single-mindedly dedicated his life towards planting the seeds of Krishna Bhakti in those souls who desire to experience the universe with Lord Krishna as its sustaining point of reference. Even in his late seventies, Shri Shyamubava is tirelessly committed to research, author and publish all identifiable and yet unpublished works of Mahaprabhu Shri Vallabhacharya.

He has also constructed a structured philosophical study curriculum and teaches an advanced course in English, Sanskrit and Hindi on "Vallabh Vedant" in the respected University of Mumbai's Philosophy Department as the penultimate expert in this Vallabh Vedant philosophy. **Shri Shyamubava has lived his life by example, as a perfect embodiment and union of all the Yogas.** For the person who wants to understand more about the entire life and his philosophical and literary body of work, the wiki link is published here. For the curious seeker, this will act as a great source of information. https://en.wikipedia.org/wiki/Shyam_Manohar_Goswami

Shri Shyamubava with his divine blessings has propelled this author on the path of continuous learning and listening to "Inner" music by daily self-practice. *Whatever knowledge, spiritual and musical qualities are evolving in this present body of this author are due to his Guru's grace and due to singing with Shri Shyamubava in his home temple at Kishangarh.* Many unknown and "Aprachalit" Ragas he makes up on the fly with his creative devotion. His musical depth is immeasurable and self-inspired by Lord Shri Krishna. *Shri Shyamubava plays Xylophone as his instrument of choice and is well versed in every aspect of the deepest Rasa of Indian music.* Music for him has its sole purpose as an offering of a loving ingredient to serve Lord Shri Krishna.

Shri Shyamubava above with his affectionate meeting on the 89th Birthday of Pandit Jasrajji.

Over the years, Shri Shyamubava has also been kind to highlight and illustrate for Pandit Jasrajji, his blessed insights on the intricacies of Bhakti centered devotional music of Pushti Sampradaya. Both Pandit Jasrajji and Shri Shyamubava share a great deal of affection for each other and Shri Shyamubava loves to listen to Shri Jasrajji sing when he is in Mumbai. Shri Shyamubava splits his time in his two seats of residence one in Mumbai at Vile Parle and other at Kishangarh, Rajasthan where he invariably spends the Diwali (Indian New Year) sharing his insights with Krishna Rasik souls there.

Even a brief moment in the company of Shri Shyamubava puts one's soul on the path towards partaking Rasa in Shri Krishna's eternal Lilas.

NL Shri Mukundray Goswamyji, Mumbai

"Veena Vadan Tatvagnya", Bada Mandir, Bhuleshwar, Mumbai

Playing his Veena above and accompanying friend Ustad Amir Khan

Shri Mukundrayji was the uncle of this author's spiritual Guru Shri Shyamubava. He represented the 15th generation lineage of Shri Vallabhacharya. His Grandfather Shri Jivanji Maharaj was a great sitar player of India having learned from Senia Gharana and the famous Sangeet Martand Prof. Bhatkhande was a music disciple of Shri Jivanji Maharaj of Bada Mandir in Mumbai. (Shri Bhatkhande performed invaluable service to Indian Classical music by formalizing and structuring the Ragas in a curriculum format in his Sangeet Shashtra volumes.) Mukundrayji was born into such a musically and spiritually inspiring environment.

Shri Mukundrayji had also spiritually blessed and initiated the author when he was in his early childhood and graced him with his musical blessings to progress towards his path. He realized that the devotional Krishna focused Pushti Sangeet (centered primarily on Dhrupad style of singing) must be formalized and documented as a reference for future seekers. Shri Mukundrayji published the *"Nada Rasa"* volume of Dhrupad and Khayal based compositions of Pushti Margiya Krishna Devotion Padas. The present author has the first edition published over four decades ago of this Nada Rasa volume. This "Bhava" emotion-laden collection of Swara compositions is the author's sole entry point and inspiration for his Vocal and Instrumental musical sojourns. Nitya Leelasth (NL) Shri Mukundrayji played his Veena in Lord Shri Krishna's Seva and his knowledge of music was self-inspired by Lord Krishna.

He was well versed in all forms of classical arts and was known in the Hindi movie industry as well. *Shri Mukundrayji was a great friend and admirer of the greatest Indian Classical music vocalist, Ustad Amir Khan.* He was friends with Naushad and other creative personalities of the time.

Author's Mridang Vidya Guru

Shri Mahantji Vishwambharnath Mishra

Shri Mahantji of Sankat Mochan Mandir, Tulsi Ghat,
Benaras (Varanasi)

Shri Mahantji hails from a great-learned family of Benaras steeped in spiritual as well as the musical heritage of the ancient city. He is the current lineage holder of the ancient 400 plus-year-old "Akhara" and temples and institutions established by great Bhakti Saint Shri Tulsidasji who wrote "Shri Ram Charit Manas" in chopai chand. The Hindu Bhakti tradition is indebted forever to Saint Shri Tulsidasji who wrote Bhasha poetry of profound excellence with Shri Rama as his center point of devotion.

Shri Mahantji presides over the famous Shri Sankat Mochan Hanumanji Temple along with the main seat of Tulsidasji at Tulsi Ghat in Benaras. He is a Ph.D. Electronics Engineer by profession and also is the Head of Department at the Benaras Hindu University (IIT) Electronics Engineering department. He also is Chairman of the Clean Ganga Foundation which is a nonprofit focused on ensuring the River Ganges and its holy waters are cleaned of all physical and biological impurities for future generations to come. *Suffice it to say that Shri Mahantji is the essence of a true Karma Yogi* and this author personally has seen him start his day at Sunrise with Maa Ganga and end his sometimes 18 hours or longer day with Shri Sankat Mochan Hanumanji without fail as his spiritual commitment- no matter how hectic or busy the other material duties.

Shri Mahantji learned Mridang (Pakhawaj) from his Grandfather "Babaji" Shri Amarnath Mishra. He continues the musical Benaras Tulsi Ghat tradition of Mridang playing with a primary spiritual focus. Shri Mahantji's family has done a great service of the Indian Cultural musical heritage by organizing and hosting two of the most pre-eminent international Indian Classical musical conferences since many decades giving encouragement to the arts. These concerts have created a much-needed forum for students and performers who have reached their pinnacle to come into contact with each other amongst a most spiritual environment. Musicians from all over India perform in these two concert series more as a form of their devotional offering to the Lord then as a professional duty.

The *Annual Sankat Mochan Hanuman Sangeet Samaroh is a weeklong all-night concert coinciding with the Hanuman Jayanti festival hosted on the premises of Sankat Mochan Hanuman temple.* This concert series is focused primarily on all forms of Indian classical music varieties except for

Dhrupad (for which there is an entirely separate concert). This usually occurs in the month of April/May yearly.

The second series is the *International Dhrupad Festival which coincides with the Shivratri festival in India and occurs usually in Feb/March timeframe.* This is a four day all night long series of Dhrupad performances from artists from all around the world on the banks of Ganges at Tulsi Ghat. This series solely focuses on all forms of Dhrupad variety of Indian Classical music and is accompanied by Mridang (Pakhawaj) rhythmic structures. The family of Shri Mahantji has been conducting this praiseworthy service of Indian music for many decades and this effort is akin to performing "Naad Yagna" in service of the Lord. The present Author and his musical journey are connected in several ways with these musical and cultural activities supported by Shri Mahantji.

Amongst all of the above duties, Shri Mahantji also finds the time with his kind grace and smile to persevere and make sincere efforts to continuously hone and refine his Mridang Vadan skills. This Mridang Vidya steeped in the Bhakti tradition of Tulsi Ghat and Gopal Mandir, he lovingly teaches and inspires the present Author from the USA and his other Shishya Shri Tetsuya Kaneko from Japan.

Present Author had been visiting Tulsi Ghat since 2005, to listen and learn at the Dhrupad Mela in Shri Mahantji's garden. Yet for some reason an opportunity never arose to meet personally or speak to Shri Mahantji. *Twelve Long years passed like the Twelve Matras of Chautaal, then the "Sama" came* and Shri Mahantji and the author met actually for the first time at the same Tulsi Ghat in 2017. He said he had been waiting. This author had been evolving in his own way for the past twelve years with more eagerness… and now the joyous Laya pilgrimage has been initiated.

41

Shri Mahantji has been ever ready to shower his love and blessings and guide the author's "Thapiya". In one of the meetings, Mahantji blessed us with his insight when the author was little restless knowing that many years have passed and there is way too much to learn and not enough time. He said with his ever-present smile *"This journey is not meant for one Life….It is a journey of many lives."* This instantly gave this author the patience and confidence to move forward with his Mridang Vadan experiments with renewed vigor.

(Saint Shri Tulsidasji was a contemporary of another great Saint Shri Surdasji who in the same era was conquering the literary heights with his loving Bhakti poetry in Brij Bhasha with Shri Krishna as his point of focus in the Pushti Sampradaya). The present author has been blessed to have the opportunity to have been in deep contact with both the Ram and Krishna Bhakti traditions.

Mahantji Sw. Shri Amarnathji Mishra (Babaji)

Grandfather of Present Mahantji, Sankat Mochan Hanuman, Tulsi Ghat, Benaras (Varanasi)

"Baba" Amarnath Mishra of Sankat Mochan Hanuman Mandir was the founding father of the spiritual Tulsi Ghat Benaras Gharana of Mridang Vadan. He was a great spiritual soul always wearing a big smile on his face as can be seen in the picture above with his joyous "Thapiya". Not only was he a musician but also a wrestler and avid supporter of old traditions of the Tulsidasji Akhara dating back more than 400 years. Babaji started and initiated the "Naad Yagna" of International Dhrupad Mela with his friends Veena Vadak Sw. Dr. Shri Lalmani Mishra and Padmashri Dr. Rajeshwar Acharya (Shishya parampara of Pandit Omkarnath Thakur). They were worried that in the winds of Khayal and modern influences of the Indian Classical music evolution, the oldest traditions of Dhrupad Ang would be lost. If Dhrupad artists were not supported and promoted by Babaji and such devoted souls, a great cultural asset could have been lost by India. Dhrupad Ang is more meditative and emphasizes the Vilambit Alap as well as Mridang Vadan due to its spiritual roots. Till this day in temples of Brij Bhoomi and many more across India the devotional music is based on this Dhrupad form of singing.

Babaji revived the old art of Dhrupad music by establishing the original roots of the International Dhrupad Mela in 1975. This auspicious "Naad Yagna" that Shri Babaji initiated is going on for almost five decades now. We wish that this effort continues forever in service of our Lord and assists the seeking souls in connecting with the Naad Brahman.

In those early days, Mridang Vadan was in even more dire conditions. Now after almost five decades of this wonderful committed effort. Mridang Vidya has revived to such an extent that the present Author born around the same time as Dhrupad Mela, is writing a book based on his experiments in Mridang Vadan and Taal Science ...residing in the USA. All of this is possible due to the blessings of spirited souls like Babaji and their love for Indian traditions and art with a devotional perspective.

Due to his spiritual role being the Mahantji of Shri Sankat Mochan Hanuman temples and Tulsi Ghat traditions, *Babaji had the music of the "Hanumant Mat" in his veins. He also had an inner urge to play the Mridang primarily with a spiritual focus. He found such a spiritually oriented teacher in Shri Mannuji. Babaji learned Mridang Vadan from Sw. Shri Mannuji Mridangacharya of Gopal Mandir, Benaras.* Sw. Shri Shrikant Mishra and Shri Manik Munde are some of the illustrious Mridang vadaks further continuing Babaji's Mridang traditions.

Padmashri Dr. Rajeshwar Acharya
"Prabhavrang"

The author is having the pleasure of knowing Padmashri Dr. Rajeshwar Acharyaji for over a decade now in the auspices of the International Dhrupad Festival co-founded by Acharyaji almost 50 years ago. Dr. Acharya has been most kind in spending his valuable time instructing this author and discussing his thoughts and his ideas about music. The author is profoundly thankful and humbled by the valuable insights presented as part of the Foreward by Dr. Acharya for his book 'Science of Melody'. The knowledge of Swar and Taal obtained by this Author from a Guru like Acharyaji is priceless. Acharyaji is a multi-faceted personality who at the age of ten years obtained Indian musical training under the guidance of Gwalior Gharana's two doyens of Indian music. Pandit Omkarnath Thakur and his disciple Padmashri Pandit Balvantrayji Bhatt "Bhavrang".

Dr. Rajeshwarji was recently recognized for his contributions to the field of Indian music by the Government of India by awarding him one of the highest professional achievement awards Padmashri. Acharyaji after completing his doctorate in music has taught many curious souls the intricacies of Indian forms of musical philosophy. In 1975 he revived the dying art of Dhrupad singing back then. Since then almost half a century has passed and through his untiring efforts in encouraging this ancient tradition, he has supported over 600 Dhrupad musicians and Mridang vadaks in the auspicious International Dhrupad Festival at Tulsi Ghat.

He has received numerous awards in his long life dedicated to Sadhana of Indian music. He not only is a great Dhrupad and Khayal singer himself but a world-renowned artist of Jal Tarang which is a rare form of art to produce music from water-filled bowls. He presided for over 24 years the music Department at DeenDayal Gorakhpur World University in Uttar Pradesh in India. He has conducted many research and scientific experiments related to music, which this author has had the pleasure of discussing with him. Through this ongoing dialogue of learning Pandit Rajeshwarji has showered his grace of knowledge on the author for which he is forever grateful. Acharyaji has always inspired this Author with his insights. One such deep perspective taught by Acharyaji will forever remain in Authors' heart;

"Music is Applied Philosophy"

This present effort to present in English the spiritual and philosophical dimensions of Swara, as well as its sensibilities, is a direct result of such inspiration and experimentation by the Author.

Sw. Shri Mannuji Mridangacharya

"Mannuji" was a Krishna devoted soul who unfailingly played Mridang during Kirtan Seva in Gopal Mandir in Benaras for Lord Krishna daily. His "Taal" was offered in the service of God. He was also the head of the Taal section teaching at the famous BHU (Benaras Hindu University). He learned Tabla and Mridang from Pandit Bholanath Pathak of Benaras. He was a very learned and selfless soul who made the rare effort to document his whole life's Taal lessons in a four-volume series called Taal Deepika in the year 1934 almost 85 years ago.

The present author was fortunate to have revived few rare remaining copies of these books from his relatives' house while they were literally rotting under a mattress. It is God's blessed Leela that this author is able to translate a portion of Shri Mannuji's effort into English in this volume and is dedicating it to the same source of all Mannuji's music, Lord Shri Krishna.

We do not have a photograph of Shri Mannuji else we would have surely published it here. The words that are flowing in this volume in English are a continuation of this great soul's devoted passion for Laya and his foresight of sharing it with others.

Sw. Mridangacharya Shri Ramshankar Pagaldas

(Pagal Baba), Ayodhya

Shri "Pagal Baba" as he was lovingly known was Pagal (Mad) in his single-minded devotion to the Taal Science and the art of playing Tabla and Mridang. He offered all of his musical abilities to Lord Shri Rama and resided in Ayodhya, where he taught many students as well. He was one of the rare Gurus who had the foresight like Shri Mannuji of Benaras, to document and synthesize his life's work in a book format for future enthusiasts like the present author. In 1965, He published his "Mridang Ank" with the complete theoretical and practical guidance for learning intricacies of Mridang from Sangeet Karyalaya, Hathras. His effort and his humility in writing the book and selflessly sharing all of his knowledge four decades ago is a testimonial to his unflinching service to

Taal Shashtra. Pagalbaba has been the author's first Mridang Guru in virtual presence through his "Mridang Ank". His soul is ever playing in front of Lord Shri Rama.

In one particular instance when discussing Taals, Pagal baba wrote in one of his notes that *"Taal Science is so deep that one Life might not be enough to master even a single Taal, let alone mastering All Taals and becoming an expert in all of the Taal Science."* This quote from him and its humility touch our hearts deeply. This author has not met Pagal Baba but our souls are indeed connected.

The flickering light of hope of the author's Mridang progress was kept alive by Shri Pagal Baba's selfless efforts in writing "Mridang Ank" many decades ago for precisely this type of inspiration for Laya seekers.

NL Shri Devakinandan Goswamyji Indore

NL Shri Devakinandanji Goswamy also represents the lineage of Mahaprabhu Shri Vallabhacharya. This author met Shri Devakinandanji, by chance at Shri Nathdwara (Rajasthan) in 2006. In the temple of Shri Vitthalnathji (a form of Shri Krishna), where he was visiting for darshan, a Kirtan Pada was being performed depicting the mood of Raas and Taal was "Charchari". Shri Devakinandanji was performing Aarti of the Lord and the author was amongst the darshanarthis doing the darshan. Their eyes met and after the darshan, Shri Devakinandanji expressed the desire to meet this author and he blessed him with his two CDs, one of Dhamar Taal and one of Chautaal. His words and blessings will be with the author forever. He had instructed in that meeting; *"Counting of the Laya on one's fingertips will alleviate all stresses of the mind and take you to a place of eternal peace."*

Shri Devakinandanji was the one of the foremost Pakhawaj vadak of India while he performed on stage publicly and in privacy for his own Lord Shri Krishna. He performed Chautaal in the International Dhrupad Festival at Tulsi Ghat in 2010. While Author and Shri Devakinandanji met only once, he left his Tulsi Ghat Chautaal recording as a milestone and guidepost for the Author. He learned Mridang from Nana Panse Gharana parampara and Shri Pannalal Pawar was one of his teachers.

'Rasik Pritam' Shri Hariray Mahaprabhu

Shri Hariray Mahaprabhu was the 5ᵗʰ generation Grandson of Mahaprabhu Shri Vallabhacharya (founder of Shuddhadvaita Krishna Bhakti Devotion path). He was a prolific musicologist and one of the most devoted authors composing hundreds of literary works and amazingly beautiful love-filled Poetries of Lord Krishna in Brij Bhasha, Sanskrit, Punjabi, Gujarati, and Hindi. He wrote most of his poetic works under the pen name of "Rasik Pritam" - (the great admirer of all the Rasa of Lord Shri Krishna.)

Shri Harirayji was intensely emotional in his approach towards Krishna. He was known for his humility and his profound devotion to his founding Guru (his ancestor and great Grandfather) Mahaprabhu Shri Vallabhacharya. All his life and efforts were dedicated to guiding himself and other souls towards the infinite Rasa in the Lotus feet of Lord Krishna. He had a long and most peacefully revered life that graced the earth for 125 years. Having read most of his philosophical works as well as his poetry, the present Author is indebted forever to Shri Harirayji for inspiring and guiding his Naad Yatra.

Science of Music - Rhythm and Naad Brahman

With blessings of Lord Shri Krishna, I hereby bow to him to allow me to introspect, experience and explain the inner sensibilities of the Science of Indian Rhythm called "Taal" for those souls who are curious to learn and are hungry for creating and partaking in the joy of the infinite musical Laya. - Author

Science Art Philosophy Religion

The subject and title of our volume is the Science of Rhythm. It is, therefore, an ideal place for us to start by understanding what purpose of this "Science" here in the context of Rhythm and Music means. Let us first explore how Science fits in with the three fundamental aspects of the universal human pursuits Philosophy, Religion, and Art.

The definition of all of the four major pursuits of life described above (Science, Art, Philosophy, and Religion) deal with mainly three aspects

1. First is the inner (physically unmanifested) universal operating laws of all creation around us,

2. Second is the outer (physically manifested) universe and creation all around us, and

3. The third is the "Creator" of these inner and outer natural laws as well as the physically manifested reality.

Science is the pursuit and thought process concerned with "Observing" and trying to "Quantify" the inner workings of Natural Laws. Science objectively tries and wishes as much as possible to "mold" these observations into our rational minds and define the "Physically manifested Natural World" around us in the confines of the "rational logic" of the mind. There is no place for "Unquantifiable" subjectivity and feelings in Science. The purpose of scientific pursuit is to identify in the rational confines of our mind the identity of who the **"creator"** of our physically manifested world is and the reasoning logic behind this creator and his creations. *Science deals with facts of our Life that can only be physically observed, measured and repeated.*

Art is the pursuit of our human existence to "Experience" the emotions and feelings of the inner laws of nature. Art tries to connect through those emotions with the creator of these natural inner truths. Art is comfortable in dealing with the "Unquantifiable" sensibilities of our human existence. Things that are not confined in rational thought of our minds are acceptable in Art. Purpose of Art is to experience the Joy experienced in the inner unmanifested natural laws and personify this Joy in the Physically manifested world around us all through a "process" that is amenable to our Hearts. The objective of Art is ultimately to identify and connect with the **"creator"** of this physically manifested world and appreciate his creations through a process likable by our Hearts and emotions. *Art deals with Sensibilities of Life.*

Philosophy is the internal analysis of our human race which is concerned with observing the Outer physically manifested realities and then making efforts to understand the inner laws of the operations of this manifested universe. Philosophy deals with the faculties of rational thought in our minds and the pursuit of producing an **"observer"** philosopher who is aware and has mastered the knowledge of the inner natural laws of all things in the Universe.

Religion is concerned with "feeling" and connecting with the inner laws of nature through a process that is amenable to our Hearts "faith". Objective of religion in this sense is to produce an **"observer"** who can appreciate and partake in the infinite Joy existing in the inner natural laws of life. The purpose of religion is to know and experience the ultimate bliss and ultimately unite with this Bliss.

As we will explore in the book, Music is one element of our human existence that is Universal in its nature. ***Music has the power to encompass all the four dimensions of human existence as discussed above (Science, Art, Philosophy, and Religion).*** Music is a means of approaching the unmanifested, manifested and the creator itself through its practice crossing all of the above four perspectives. We are striving in this volume to explore some of the Science (quantifiable logical aspects) and Artistic sensibilities (emotional) underpinnings of the Rhythmic components of the Indian Classical Musical system. As we analyze the Rhythmic Science of Indian Taal Shastra Author's frame of reference is certainly steeped in

Philosophical and Hindu Vedantin religious spirit.

Music - Philosophical Evolution

All forms of musical systems in India, be it Vocal, Instrumental (non-rhythmic) or Rhythmic, are derived from the root of "Naad Brahman". So before we get to Rhythmic Taal Science and its structure, let us explore the philosophical meaning of this Naad Brahma. "Naad" means Primordial sound in its commonly understood English meaning. However, this Naad in the Indian system is not just sound but in our Philosophy the Naad = Brahman itself (the all-encompassing universal Creative and evolutionary force also called Par-Brahman). The Brahman in the Indian Vedic system takes infinite forms and proceeds to evolve into many worlds and many realities of which the Naad is one aspect of the infinite aspects of Brahman itself.

Now we might ask why and how is this Naad created. The Indian Philosophical system of evolution explains it in the forms of Aahat and Anahat Naad, which we shall explore just in a while. Both the forms of Naad mentioned here start with the root Sanskrit word "A".

Anahat Naad

To study the genesis of Naad we must go to the roots of the Indian Vedic System and its concept of evolution. For this, we reference Vedic Brahma Sutras *(codified by the sage Shri Ved Vyas who also wrote Shrimad Bhagwat as well the Mahabharata)* that highlight the system of *"Brahma Vada" (the philosophy pertaining to understanding and experiencing the ALL permeating creator Par-Brahman)*. These Brahman principles were documented by the Sage Ved Vyas in his

57

Brahma Sutras. Per "Brahma Vada", this Brahman is first existing itself as it is all-permeating and ever-expanding. It then initiates the internal dialogue and debate about its need to "create' the "Shrishti" (our present world) and all its corresponding paraphernalia. This inner dialogue of Brahma with its own self is called "Aabhu". **This internal "Aabhu" dialogue of Brahman with its own self is nothing but "Anahat Naad".**

This Anahat Naad is ever-present in the entire creation and is essentially Brahman itself since it's the inner Dialogue of Brahman and is essential for the creation and existence of the Worlds around us. "Worlds" because our world is not the only one and our sound is not the only one. However, this Anahat Naad is not "lok ranjak", meaning it can't be heard by ordinary humans and its main purpose is not to produce Joy in humans. *The main purpose of Anahat Naad is to act as a foundation for all of the universe.* Only the souls that have attained oneness with Par Brahman and the sages and Rishis of the yore could experience and hear HIS vibrations of the Anahat Naad.

Aahat Naad

As described, the inner dialogue of Brahman with itself "Aabhu" occurs and the corresponding "Anahat Naad" is produced from this inner dialogue. However, in our Naad Brahma philosophy, there is the same universal concept of the male and female (or yin and yang). Brahman now wishes and decides to multiply in its infinite forms for its own Joy. And as it manifests in its infinite multitudes of forms, Brahman now starts conducting the debate and conversations with its own infinitely multiplied evolutionary forms. This dialogue is called "Aa Bhuvanti".

This is the dialogue in the many realms of the Brahman that it conducts with its own multitudes of forms subsequent to their evolution. One such dialogue of Brahman happens in our present Physical realm that we live in. **The sound produced from this Brahman conversation "Aa Bhuvanti" with its own physical multiplicity of forms, is nothing but what we call the "Aahat Naad".**

This Aahat Naad is the one that we humans can hear and this is the Naad that is "lok ranjak", meaning we can hear it, feel it, experience it and enjoy it in its form as sound and music. So to summarize the concept, the physical multiplicity of forms of Brahman starts conversing amongst itself and thus produces this "Aahat Naad". In the musical sense, Aahat Naad is comprising of the seven notes that we hear 7 Swara and the myriads of Rhythmic intonations and other grammatical sounds and intonations that are part of this Naad. We know this as our Music. This Aahat Naad (Music) is a universal concept but across the world, the physical realms of Brahman's creations manifest in different forms of different religions, races, and countries, but the concept of Music remains universal common denominator unifying us all.

Joy in Music – Philosophical and Scientific "Resonance"

Since time immemorial, the philosophers and sages across the world have pursued their quest for understanding this Brahman (known by many names and many forms at different times, places and regions). To seek it and to be one with it. This same concept of Oneness is the internal purpose of all music whether it is knowingly pursued by the musician after studying the roots of this philosophical connection with Brahman or unknowingly pursued through just the act of singing or playing for Joy.

The fundamental scientific principle of "Resonance" is at work here. We know that when there are two different frequencies occurring close by to each other, the closer they come in their physical vibrating range, the closer they come to a point of "Unified Vibration" also known as the point of "Resonance" This is the point when the frequency gets amplified and the two separate frequencies become 'ONE' producing a single unified frequency.

When a musician or listener says that they feel they connected with the Spirit of God or the creator, what they are essentially saying is that their "Aahat Naad" is finding a moment of "Resonance" with "Anahat Naad", the physically manifested and the all-pervading spiritual Naad Brahma unite and that is the climax of Music. A commoner feels that Music has this power but the common man does not know the science and philosophy of this Naad Brahma and its resonance. He/she can feel it when singing or listening and that is the Joy and Beauty of Music and its Universality.

This **_Resonance_** between Aahat Naad (music that we can hear) and Anahat Naad (the music of Brahman's internal dialogue) is the ultimate objective of Oneness. A person singing, who with years and (often many a lifetime's) practices arrives at the perfect resonance with one of the seven musical pitches, he/she sometimes experiences roots of the hairs on skin tingled, or they have tears in their eyes. What is happening is that the singer is feeling this moment of oneness with the Brahman (the Creator).

Musicians also know this joy in arriving at the point of **_"Sama"_** that we shall explore later. This Sama is a representation of this Resonance between Rhythmic and Non-Rhythmic cycles of music. The point of Union and Resonance and its Joy is infinite and impossible to describe in words, just as the infinite Brahman can't be described in our human words but can only be felt for the few who unflinchingly pursue this "Oneness" in the form of Naad Brahma

So this is Naad in its elemental "Aahat" form that is heard by us. Naad in musical form that we hear is briefly of two formats;

1. Swara (connected with Vocal singing) – the 7 Notes comprising of their corresponding pitches and scales, etc. used in Vocal music as well as Instrumental (non-rhythmic) Music. Swara is used in singing or playing non-rhythmic instruments (Melody oriented). As the present volume is focused on Taal we will limit the "Swara" discussion to this brief definition. Swara and Melody will be covered in-depth in a separate volume to be published later.

2. Laya (connected with Taal and its rhythmic beats) - The myriads of rhythmic sounds produced by the rhythmic instruments. Laya is used in the creation and execution of the Taal and we will define these ideas now in the book.

   These beats, the science and art of producing those sounds and using the rhythmic instruments are the subject of the present volume and this is called the "Taal Shastra".

Naad Psychosomatic production.

Having understood the Philosophical background of Music, Taal, and Swara, we now explore the precise inner psychological process as well as the physiological process in producing the sounds of rhythm. This process is described in much details in the Indian Vedic scriptures and Upanishads, but we summarize it here to understand our subject in its entirety. This process and its execution with spiritual goals is the starting point of Naad Yoga to be introduced separately in the chapter.

(The Psychosomatic process of Aahat Naad creation is as explained below in its true natural spiritual science originally described in Sharangdev's Sangeet Ratnakar treatise)

'आत्मा विवक्षमाणोयं मनः प्रेरयते मनः । देहस्थं वह्निमाहन्ति स प्रेरयति मारुतम् ॥
ब्रह्मग्रन्थिस्थितः सोय क्रमादूर्ध्वपथे चरन् । नाभिहृत्कण्ठमूर्धास्येष्वाविर्भावयति ध्वनिम् ॥'
'नादोतिसूक्ष्मः सूक्ष्मश्च पुष्टोऽपुष्टश्च कृत्रिमः । इति पञ्चविधा धत्ते पञ्चस्थानस्थितः क्रमात् ॥
अपवहारे त्वसौ त्रेधा हृदि मन्द्रोऽभिधीयते । कण्ठे मध्यो मूर्ध्नि तारो द्विगुणश्रोत्तरोत्तरः ॥
तस्य द्वाविंशतिर्भेदाः श्रवणाच्छ्रुतयो मताः ॥'

1. First, as we described above, the Brahman residing in our "Souls" wishes to have a dialogue and hence wishes to produce some sound.
2. At this point, the Soul (Atman) triggers our mind for its wish to produce the sound.
3. The Mind in turn "Ignites' the fire energy residing in our physical human body.
4. The inner "Fire" energy in our Body activates the "Vayu" element of Air in our body now.
5. The "Vayu" resides in our Brahma Granthi (the Perineal Knot in our yogic body).
6. This "Vayu" when activated by "Fire" moves up and rises from Brahma Granthi in the following order of ascent, to

our "Naabhi" (Navel), then "Hriday" (Heart), "Kanth" (Throat), "Mastak" (Head) and finally "Mukh" (Mouth).

7. Sound (Aahat Naad) is produced by "Vayu" vibrations moving up in all the five locations described above.

8. This Naad produced thus is categorized as below;

    i. Naabhi (Navel) - Ati Sukshma (extremely micro)
    ii. Hriday (Heart) – Sukshma Naad (small sound)
    iii. Kanth (Throat) – Pusht Naad (fuller sound)
    iv. Mastak (Head) – Apusht Naad (thinner sound)
    v. Mukh (Mouth) – Krutrim Naad (artificial sound)

This is the scientific and spiritual root of sound production in human beings through which music is produced.

The sound produced thus evolves into Music in the following manner.

    i.   From the Hriday (Heart) "Mandra" Naad is produced (Lower Octave) – (Sukshma Naad)
    ii.   From the Kanth (Throat) "Madhya" Naad is produced (Middle Octave) – (Pusht Naad)
    iii.   From the Mastak (Head) "Taar" Naad is produced (Higher Octave) – (Apusht Naad)

This "Aahat Naad " as it evolves, increases with double intensity in a larger and larger higher pitch and covers 22 pitches. This is how music and the sounds are produced from our physical bodies. This is the inner deep science. This is THE spiritual-scientific process. The Swara and Taal (rhythmic) sounds are produced in the same fashion. In the case of either type of the Swara or Taal they can be produced in a solo manner by the above process, or in a combined manner together in which case Swara is produced and Taal and its pronunciations and time boundaries are superimposed on it. **Their pronunciations are different and the techniques of reproducing them physically outside of our bodies on different instruments are different and that's where the entire science migrates from being Psychosomatic to the physical realm.**

Science of Rhythm

Expanding upon the above Vedic process of Naad creation one can now analyze the production of the Rhythm. In Step 8 above when the soul and body decide to produce a specific Rhythmic sound now following extra steps happen;

9. Based on the type of Naad, the rhythmic sound through the activation of internal fire and air comes from one of the specific five locations in the Body as discussed above;
10. The mind decides based on the soul's wish, if this rhythmic sound is to be pronounced from the human body verbally (then the rhythmic sound comes out with effort from one's verbal speech of mouth)
11. If the rhythm is not to be verbalized with mouth then the sound is internalized in mind itself but the rhythm is indeed "Spoken" in mind first.
12. Now if the soul desires to extend this process to its physical realm it wishes rhythmic intonation to be "reproduced" on an external instrument.

13. If this is the case then the mind triggers the nervous system to activate the necessary limbic response of (in case of Mridang or Tabla player) the hands and/ or palms and fingers.
14. Based on the "emotion" desired by the soul, the mind decides if the limbic response intensity is to be soft, hard or varied.
15. Once decided and based on past feedback from repeated lessons and practice, mind and body have been trained to decide to "Strike" accordingly and act upon another surface to reproduce some rhythmic tone.

Lok Ranjak and Atma Ranjak Aahat Naad

We have thus explored the process of manifestation of the Aahat Naad from a complete Spiritual, Psychological and Physical perspective. Let us understand the two ends of the spectrum of Aahat Naad in the context of a Musical performance. We have already discussed that Aahat Naad is "Lok Ranjak" meaning it is the one which gives Joy to humans as they can hear it. But two categories of listeners are there in performance. The ones who are performing (musicians themselves) and the audience are the two categories so in that context "Lok Ranjak" means the music that is enjoyed by the audience. This is one end of the spectrum of Aahat Naad.

But the performer is also hearing. **For that performer what is more important in Aahat Naad is what we call "Atma Ranjak" which is the other spectrum of Aahat Naad. What we mean by this new term is that the performed music is felt and produces Joy in the "Antahkarana"** (All faculties of the performer with the soul – Atman at its center). This is the unique personal experience of "Atma Ranjakta" for the performer.

What we see in most Indian Classical Rhythmic, Vocal, or Instrumental performances is by and large following structure in a very simplified fashion.

1. Alap (Initial phase) of performance in Slower tempo with more use of Lower Octaves and no beats. This is just a melody. As we have seen the Lower Octave is produced more from the Heart in the process of Naad creation. This section of the performance is naturally more "Atma Ranjak" as the Musician is able to go deep into the melody or slower tempo of Rhythm if it's a solo Rhythmic performance. This also is a very difficult aspect of performing. The modern Audience who are less patient can be turned off if they are not familiar with the process of what is happening and this Alap phase therefore is less Lok Ranjak for a modern audience. But without a doubt, this is the phase which is intensely "Atma Ranjak" for the performer.

2. Madhyam Vistaar (Middle phase) - In this phase of performance now the musician is expanding the mood that is established from the Heart and moving more to the Throat (Kanth) Middle Octave of melody and also Rhythm is introduced and a lot of permutations and combinations of Swara and Laya can be produced here. This section has the possibility of being more Lok Ranjak in the Aahat Naad spectrum of Joy for the audience. And depending on the harmony between melody and rhythm it can be equally "Atma Ranjak" but not as much as in the previous phase. Hence this is the balance and in the middle of the Ranjakta scale between the two ends, we have defined.

3. Drut (Taar) Vistaar - The final and the third stage of performance happens in what we call Drut or Fast tempo in Rhythm and the Higher Octave being deployed in Melody. As we have seen earlier the Taar Saptak (Higher Octave) uses sound originating more from Head and it's called Apusht Naad. This is more a phase of "Physical Exercise" to demonstrate the physical skill level and prowess of the performer. This is where showmanship comes first and the Audience now is on full alert as they don't need to be patient just to hear with their ears but they can see the actual physical form of the performance as well and the tempo is fast. The faster the better. This is the third stage which is mostly defined for its "Lok Ranjak" ability for producing maximum joy in the Audience that they will remember. But this is phase is not as useful for emphasizing the inner faculties of "Antahkarana" and has the lesser possibility of "Atma Ranjak" joy in the soul. The Joy here is fleeting and the tempo is fast. Here the physical prowess is to be measured not the internal depth of the soul and Naad being produced as in the Alap.

1. Now when a performer has reached the pinnacle of their art and has understood its deeper science as well. They can achieve what few can achieve which is to make the Drut section of performance reach such a stage that it feels and sounds and senses like Vilambit. This is the ultimate prowess in completing the feedback loop of Atma Ranjakta and Lok Ranjakta. This is the ideal that any enlightened performer must strive to attain. **This loop completes the infinite feedback of Joy.**

Thus we have seen that even amongst the Aahat Naad there is a spectrum of Joy that evolves. What is seen is that the more blessed and seasoned the performer, the more they have the ability to come in the middle of the Aahat Naad spectrum where the Naad is equally Atma Ranjak as well as Lok Ranjak across all the three phases of evolving performance. However, the fact is that "Lok Ranjakta" is the main measure of success in performance if the audience and their claps are the metrics of measurement. To do things that please the audience many times the performer sacrifices their inner Joy in Naad creation (Atma Ranjakta).

The performers at the pinnacle of their art described in step 4 above are those that do not deviate from their main focus which is their own inner Joy and sing or perform rhythm for their own soul. This "Atma Ranjak" feeling then *radiates* from the performer to the audience and the audience is also captivated in the process even if not as knowledgeable in the actual act of creation of the Naad. *The Souls of the Audience are now touched by the Joy of Atman of the performer and they are also able to experience "Atma Ranjakta". In those rare moments, the Aahat Naad reaches its pinnacle of resonance between the Performer and Audience. They become one. They both are listeners' with equal Joy of Atman. This is the Objective. This is the "Saarthak" performance.*

Naad Yoga and Taal Shastra

"Yoga" has been the gift to mankind coming from the deepest origins of Indian philosophy and the Hindu Vedas. There are many different types of Yoga as having been described by Lord Krishna himself in his Bhagwad Gita addressing the warrior Arjuna in Mahabharata. Few Yoga examples are listed here. This list is not all-inclusive as there are many paths of Yoga besides these few examples.

1. Karma Yoga – Action as the instrument
2. Bhakti Yoga – Devotion as an instrument
3. Gyan Yoga – Knowledge as an instrument
4. Hatha Yoga – Intense physical control
5. Naad Yoga - Music as the instrument

The type of Yoga that interests the seeker of Rhythm here is "Naad Yoga". The term Yoga in general in Sanskrit means "to yoke" or "to join". We have already described so far the "Evolutionary" aspect of the Brahman in Indian philosophy. And this Brahman is the one who amongst other things is the creator of the Naad – the Primordial Sound of the universe and thus the Naad = Brahman itself. All Yoga practices aspire to ultimately lead to the awareness and union with the ultimate reality = Brahman. In Naad Yoga one uses the format of Naad and its processes and science that we have summarized earlier to unite with the Creator (Brahman) and his many aspects of realities all leading to itself.

All Yogic practices have a prescribed format and structure within which one must execute these to achieve some desired results. **The Naad Yoga in its simplest form is the practice of Music with a Spiritual objective in mind. This Musical practice could be any form of Music (Vocal, Melody or Rhythm or even Dance) so long as it has the oneness with the *Brahman* as its objective.** The practice essentially is a process of becoming aware of the universe around us gradually appreciating it with its many variations and ranges with Naad "Sound" as the entry point. Those universal infinite possibilities take manifestations in forms of Swara and Laya becoming "alive" in the experiential sense for the Naad Yoga practitioner. As in all Yoga practices that are focused inwards, the practice is not a social one with many people but rather is best performed in front of one's spiritual point of focus. For this reason, almost all professional Musical performances are not suited for experiencing true Naad Yoga. By its very definition, the practice and experience of Naad Yoga occur in the individual pursuit and is not a collective experience.

A student of Naad Yoga almost invariably knowingly or by chance ends up aligning oneself also with Bhakti Yoga in parallel. Since the earliest Indian system of time, it has been indicated in Vedas that during the present Indian cycle of time "Kaliyuga", **"Singing" the music in loving praise of the Creator and his many creations is the ultimate pathway to attaining oneness with him.** This message of Bhakti Yoga and its emphasis on loving the Lord and his creations with singing and music aligns itself well to the Naad Yoga practice as well. So essentially a person being a student of Bhakti Yoga might not necessarily be a student of Music or Naad Yoga. But the reality in other way is clear that a seeker who is walking and progressing along the path of Naad Yoga almost invariably will end up also being familiar with and become an appreciative practitioner of Bhakti Yoga.

Mindful Meditation in Music

The above Oneness with Brahman through the use of Naad as the medium is easier said than done. It takes years and many births to achieve this type of success. However, a true seeker is not daunted by all this and "Patience" is his/her biggest virtue of success. So when one practices Music with "Conscious" understanding and "Deliberate" focus on this spiritual objective of Naad Yoga, what occurs is that the "Mind" which is ficklest of all our faculties slowly begins to attain more and more consciousness of first the "Aural" universe around us and then through that frame of reference other aspects of life as well.

Said, in other words, Mind slowly begins to enter what this author calls "Mindful Meditative State" - in Sanskrit this term is called "Avadhaan". This Avadhaan is not the usual concentration in its normal singular sense of focusing on one thing. Rather the Musician naturally progresses towards true *Avadhaan whose definition is "Parallel Processing and Parallel concentration"* without losing sight of the individual details. So we can appreciate that this Mindful Musical Meditation is unlike the process of the more common definition of meditation which in some practices tries to "Empty" the mind and become an observer etc. This process here in Naad Yoga is inverse. It is to "FILL" the mind with the Swara and Laya and their inner psychosomatic possibilities. Concentrate in Parallel on multiple layers of Musical "Naad" and its multiplicities and nuances. Striving to achieve perfect "Resonance" with each note of Swara being executed upon or each Beat of Taal Rhythm and slowly the mind learns to attain the "Mindful" state of "Avadhaan". This is the state of mind where it is all "Aware" and "In the Present moment it is one hundred percent committed." Free from any distractions and totally in oneness with multitudes of possibilities within the domains of Swara and Laya. This ends up being the Mindful "Avadhaan" of the Naad Yoga practitioner. It does not happen on the first day but with constant committed and deliberate exercise of sitting down and practicing Music, our mind gravitates towards more and more state of Avadhaan. Now let us put together all the ideas we have discussed from the scientific and spiritual side and analyze what happens practically.

Reverse Engineering Musical Process

Deconstructing the Act of Singing - When one sits down with a Harmonium (keyboard type of instrument) used for Swara accompaniment with Vocals. Here are the practical things happening at the "Mindful" level in each such experiment of a singing session.

1. A Tanpura (Drone) is initiated in the background to produce the "reference notes" of Shadaj (Sa) or in western scale (C) notes.
2. The singer's mind now listens and gravitates towards using that reference note as a base to get into the mood of the scale of singing.
3. Then the mind acts upon all the instructions it receives from the soul as we have discussed earlier to decide which Swara it needs to produce.
4. The mind decides that the soul has been instructed to sing so the Mouth starts producing the sound.
5. Simultaneously or a few 1/10ᵗʰ of seconds prior to this the mind instructs the fingers of the Musician to pump air in the Harmonium and press the corresponding scale key in Harmonium so that the act of "resonance" can be created between the singers Vocal Chord produced Swara note and the Harmonium note.
6. Now that's not the end of it all, the Soul desires to listen and actually hear the sound of sung "poetry" praising the Lord. The poetry or Kirtan has its own grammar and words and the mind "superimposes" these words on the Swara note that is being sung.
7. The actual Swara is silently being heard in mind through its past practice of reference tonality but the words of poetry are now superimposed in that Swara with "resonance"

being facilitated by usage of Harmonium playing and Tanpura.

8. Now you add another layer of complexity to this if it's not already complicated. The above process is to be "bound" in a cyclical beat timing of a specific Taal cycle and so the mind must calculate and have a practice to "Start" and "Stop" each new line of poetry being sung with a cycle of the repeating Taal loop.

9. The Musician's mind then is also simultaneously thinking ahead to decide the next sequences or paths of notes and words and beats to combine to produce the next section of singing music. Each section is like an "inner pilgrimage" from one specific place "Makaam" (in Urdu) of Swara and Laya to another and back with many exits and entry points.

10. This process repeats in fractions of a second inside the mind of a Musician – first, the sound is originated in mind, reproduced in instruments which can be our vocal chords or the string, wind or rhythm instruments. The resonance is to be found, in each inner step to practically perform the act of singing.

Act of Rhythmic Taal playing - Similarly many parallel processing activities are taking place in the mind even when one is playing the Rhythmic instrument or just an Instrument like Flute. In the instance of Rhythm all above steps are practically occurring with one main difference that in the minds of the Rhythm player the fixation is not the pronunciation of the Swara but the "Bol" or the "Varna" sound of the Rhythmic strike and he/she either verbalizes that through the mouth if playing solo and simultaneously produces that same sound on the Mridang or Tabla. Or in case of vocal accompaniment the Rhythm player keeps their verbal sound silent and produces the Bol only in mind and just the rhythmic instrument sound is heard.

Act of Instrument playing (Flute) – Here again same steps described above take place except for the fact that the Musician playing Flute has the mind fixated on Swara aspect but is focusing on physically playing the Swara on flute with his/her fingers and breath control (and not singing). That is the main difference - other aspects of mental activities are similar. The flute is an example of an instrument where it is impossible to sing while playing and hence we have used this example here (besides the fact that this author is a student of Flute as well). Some instruments like Sarangi and Harmonium are used to specifically accompany Vocal and hence the player of those instruments can easily sing as well. Plucked instruments like Veena also can accompany and are easy to sing with as they can produce all the vocal pitches.

This illustrates one thing that the Music we hear has many moving internal scientific and psychological and physical components required in its production. These all need to work like a Gearbox in an engine to create "unison" and good resonance. Even if one piece above is off by some small fraction of time, the Music is "Not in Tune" and the singer and listener both can experience this moment which is akin to death for the experienced musical soul.

Music demands a lot from its practitioner and it needs not only full cognitive maturity and awareness but the motor skills (of our limbs used to execute and produce the music) must be also fully mature. This is where "Riyaz" - "Practice Practice Practice" comes into play. The Motor skills that we have many times are also referenced as "Muscle memory" and our muscles also have awareness and Music unlocks this through the continuous practice of Avadhaan. Initially, when one start learning to sing or play an instrument, the mind is more driven towards ensuring motor skills are able to perform and the Music is produced with much effort. Ultimately the stage is reached where fingers and hands know what to do naturally

74

and the mind is free to make other creative decisions in the act of Music-making. Such perfection in the practical act of singing is desired (and that is why committed practice is a must) but is impossible to do with our mind literally floating everywhere as is the habit of every mind. *Music forces the Mind to be "Mindful" and the "Muscles to be Aware."*

A musician who wants to produce good music naturally must find the "Avadhaan" with the notes and beats first. This is a prerequisite to producing "In Tune" music. And over time as the soul and mind's practice improves, the "resonance" is to be desired with the Brahman itself. This is the reason while we have many classical musicians performing, but rarely and occasionally there are performances where a musician and his/her mind and soul unite and connect to the creator through full "Avadhaan" touching the listeners' soul directly.

Achieving this "Mindful Music" is not easy and it is rare to hear it in everyday listening. But if one has introspected and studied what the Scientific and Spiritual process as described earlier looks like and what to look for then gradually one can experience it in a lifelong pursuit as a student of Music. *A listener can appreciate the rare moment when the singer and listener unite in their soul. At that moment Naad Yoga has come to its physical fruition and arrived at Mindful Meditation.*

Neuroscience and Rhythm

The simple act of singing or playing a Rhythmic Instrument or a Flute is not so "SIMPLE" when we now understand the many layers of spiritual, scientific, and physical actions happening simultaneously.

In 2016, this Author witnessed an interview of one of the most famous contemporary western popular musicians–Rock Music legend Sting. This interview was remarkable because for the first time in human history our science had the capability to study under an fMRI the brain and mind of a lifelong musician in real-time. Sting had allowed neuroscientists, Levitin and Grafton to study in a lab his brain activity while connected with probes of an fMRI machine. Using the latest neuroscientific techniques the brain activity of one of the most successful musical geniuses in the popular contemporary art scene was analyzed while he listened to many differing playlists of music and executed the processes of creating pieces of music in his mind. What Sting said in that interview, in the end, will register with the author forever and will be shared soon. First, let us understand the observations of neuroscientist's analysis of Music.

So far, we have seen a brief but process-driven perspective of the Spiritual and Physiological extension of Naad Brahma and its aspects of Musical sound. The actual process if analyzed from a Neurological scientific lens is equally complex with many parallel neural networks being triggered and initiating many cognitive, physical, auditory as well as other sensory feedback loops. For a Musician like Sting, (a lifelong student of the art), the above study observed the neural underpinnings of music perception and cognition, in particular, musician's mental representations for music.

1. Musical training and performance practiced over a long period of time were observed by this study to leverage neural networks that were separate and distinguishable from the composition of prose or visual art.
2. Also observed was that listening to and imagining music shared overlapping neural representations.

These neural brain networks were grouped in clusters based on the key (Swara pitch and octave), tempo (Laya and Taal structures), motif (emotional aspects), and orchestration (collaborative aspects of Rhythm with non-rhythmic instruments).

Therefore, inside our brain, there are tremendous amounts of parallel activity going on within the different neural network clusters just in playing the Rhythm. (Refer to our discussion on Avadhaan earlier in Mindful Meditation and Music section). What we also find here is that the Student of Music learns in varied manners by listening as well as practicing and even imagining music and these acts leverage overlapping neural networks. The portions of the brain triggered by Music additionally were found to be different from the other artistic clusters of Poetry, or Visual arts, etc. Music triggers new and different types of brain clusters that are not the same as the ones our brain uses for Painting, Reading and other creative activity.

Another set of scientists studied the *Motor skills in the physical body and their correlation with Rhythm specifically*. Ross, Iversen, and Balasubramaniam in the same Neuroscience Journal issue on Music reinforced the role of the motor system in perceiving rhythms with a strong beat, even when the listener is sitting still. *They observed that the motor system plays a more critical causal role in beat perception. Thus underscoring potential connections between music and speech processing.* This is nothing but verifying for us from our Aahat Naad creation discussion from Vedic Science, that the music and Rhythm happen first in mind, where based on specific instructions of mind, the music is verbalized either internally or externally (through the mouth) and then the motor system takes over the process under instructions from the brain in case of rhythm and strikes and movements of limbic systems.

Now if you refer back to the Vedic descriptions of Musical process creation described earlier you will see how deep our Sages knew the inner nature of nature and precisely documented the processes of creation and execution of Music and Rhythm. The Vedic scientific approach goes much deeper and further in the process and actually explores the roots of music creation by studying the soul and Naad Brahman as well. Modern science is only catching up now and is only scratching the surface of physical and mental aspects known to the Indian Classical Music practitioners and Sages for thousands of years. Remember the subjects of Soul, Spirit, and Emotions are incompatible with the modern scientific approach as we discussed at the very beginning of this Scientific and Artistic analysis of the Rhythmic and Musical science of India.

The Neuroscientific research on Music is ongoing and we are sure to see it produce further proof of the amazing foresight of Sages of India in codifying the Musical systems as they have. This is the point, which brings the author back to what Sting had to say at the end of that interview on TV where his brain was connected to fMRI and *he was able to see literally his brain waves (no pun intended).* The Doctor asked him how he felt, seeing the activity inside his Musical brain actually being shown on the TV monitor. Sting's answered that "He is excited to see science evolve and is happy to help in that regard but he would like to switch off the monitor as soon as he can as too much emphasis on the quantification of the brain networks he feared would take a crucial element out of his creative process. **He would prefer to Feel his Music rather than Measure it in his Brain."** This was the gist of his answer as a lifelong music practitioner.

A Musician who understands the above scientific and spiritual processes described so far will have a better chance of understanding the pursuit of Naad Brahma. If one practices

Music with the deepest understanding of the ideas described here, such a person would have infinite possibilities and wonderful opportunities of progressing on the path of Naad Yoga. What is required is an intense commitment and spiritual thirst to go to the deepest possible depths of the infinitely deep "Ocean of Laya and Swara". In that journey itself, one finds the point of unification with each note and each beat created by the Brahman.

Section 2 - Ten Prana of Taal

तालों के दश प्राण।
कालमार्गक्रियाङ्गानि ग्रहजातिकला लय:।
यति प्रस्तारकौ चेति ताल प्राणा: दश स्मृता:।।७।।
भावार्थ : काल, मार्ग, क्रिया, अङ्ग, ग्रह, जाति, कला, लय, यति और प्रस्तार यह
तालों के दश प्राण हैं।

The ancient creators and sages of Indian musical aesthetics, considered the Taal to be composed of ten key elements and they considered these ten qualities as the essential "Life-force" of Taal. We can better appreciate the context of these Ten characteristics of Taal once we have some basic background of the term "Prana" in a yogic sense.

The Indian concept of Yoga considers "Prana", the life force (the breath in its material form), as the fundamental root of the principles of controlling one's own physiology and spiritual progress. "Prana" is used as a centerpiece in the Yogic techniques. **The Indian musical approach in its deepest spiritual sense is also termed as "Naad Yoga" - this music is the Naad, the primordial sound in its most elementary being, which we have discussed earlier in a separate section on Naad Yoga.** The practice of this Music is akin to Yogic exercises as it is another means of improving one's journey of physical, mental and spiritual well-being. Allowing one to progress on one's path of exploration and understanding of Naad Brahma – the "Aural" manifestation of the Creator "Brahma".

In Indian music, the Taal becomes the Prana of the music as the Taal determines based on its speed (just like how fast one inhales or exhales determines their state of physical well-being) determines the state of other aspects of music. The

Taal in its fundamental form as defined and documented by the Sages has Ten distinct "Prana" as per the Indian ancient musical texts. They are as follows;

1. Kaal
2. Maarg
3. Kriya
4. Ang
5. Graha
6. Jati
7. Kala
8. Laya
9. Yati
10. Prastar

Next we shall explore each Prana of Taal and try and get much better acquainted with them. Knowledge of these ten Prana is essential to the seeker of Indian Music and practicing student of Indian Rhythmic structure in understanding the scientific and emotional dimensions of Taal in its most natural form.

The Prana Table

(Summary of the Ten Prana for handy reference)

No.	Prana Name	Macro Grouping	Sub-Category	
1	Kaal (10)	Margiya (Gandharva)	EIGHT "Kshan"	= 1 "Lav"
			EIGHT "Lav"	= 1 "Kashta"
			EIGHT "Kashta"	= 1 "Nimesh"
		Deshiya	EIGHT "Nimesh"	= 1"Kala"
			TWO "Kala"	= 1 Chaturbhag
		(Humans)	TWO "Chaturbhag"	= 1 "Anu"
			TWO "Anu"	*= 1 "Drut"*
			TWO "Drut"	*= 1 "Laghu"*
			TWO "Laghu"	*= 1 "Guru"*

			THREE *"Laghu"*	= 1 *"Plut"*
2	Maarg (4)	Dhruv= 1 Kala Chitra= 2 Kala Vaartik= 4 Kala Dakshin= 8 Kala		
3	Kriya (8)	Nishabd Shashabd	Nishabd - Aavap Nishabd – Nishkaam Nishabd - Vikshep Nishabd - Praveshak Shashabd - Dhruva Shashabd - Shampa Shashabd - Taal Shashabd - Sannipat	
4	Ang (7)	Anudrut Drut Drutviram Laghu Laghuviram Guru Plut		
5	Graha (4)	Sama Atit Anagat Visham		
6	Jati (5)	Chaturastra= 4 Matra Triyastra= 3 Matra Khand= 5 Matra Mishra= 7 Matra Sankirna= 9 Matra		
7	Kala (8)	Dhruvka Sarpini Krishna Padmini Visarjita Vikshipta Pataka Patita		
8	Laya	Vilambit	Drut Drut Madhyam	

	(3)	Madhyam	Drut Vilambit
			Madhyam
		Drut	Madhyam Drut
			Madhyam Vilambit
			Vilambit
			Vilambit Madhyam
			Vilambit Drut
9	Yati (5)	Sama Strotovaha Mridanga Pipilika Gopucha	
10	Prastar (2)	As described two processes in the book.	

ताल के दस प्राण

कालोमार्गः क्रियांगानि ग्रहोजातिः कलालयः ।
यतिः प्रस्तार कथ्येति ताल प्राणादशस्मृतः ॥

अर्थात्—तालों के दस प्राण शास्त्रों ने माने हैं। १. काल, २. मार्ग, ३. क्रिया
४. अंग, ५. ग्रह, ६. जाति, ७. कला, ८. लय, ९. यति और १०. प्रस्तार । ताल के
दस प्राणों का ज्ञान परमावश्यक है, क्योंकि पूर्वाचार्यों ने स्पष्ट कहा है कि—

दसप्राणादिमर्कं तालं यो जानाति स तत्त्वविद् ।

(भरत-भंडारी)

अर्थात्—प्रत्येक तालज्ञ को तालों के १० प्राण तत्त्वों का ज्ञान होना
आवश्यक है ।

Taal and Its Nature

So far we have now understood Rhythm and its process in overall Music making from holistic Scientific, Spiritual and Philosophical perspectives. This prepares us to now dive deeper into the "Science of Taal Rhythm" in ensuing chapters. What you will see is that our Indian sages profoundly created the systems of Taal Shastra based on logical and scientific techniques while simultaneously enhancing the Sensibilities (Emotional "Bhavatmak") aspects of the Art. *A seeker of Taal must understand that this is a journey of multiple lifetimes and the ocean is too deep for one lifetime.*

Let us first understand who the founding fathers of this system are. In the Indian Spiritual sense, Music has several ancient originating main Schools of approach dating back to the Creator. *The original "Acharyas" of Indian Musical knowledge are considered the founding creators of the Science of Rhythm* in Indian Musical context. They are named as follows;

संगीतशास्त्र को जानने वाले आचार्यों के नाम।
सदाशिवो हरिर्ब्रह्मा भरत: कश्यपो मुनि:।
हनूमानङ्गदश्चैव नारदस्तुम्बुरुस्तथा।।४।।
दुर्गाशक्तिर्मतङ्गश्च यास्कशार्दूलकोहल:।
एते संगीतसर्वज्ञा: बुधास्तालान् प्रचक्रमु:।।५।।
भावार्थ : शिव, ब्रह्मा, विष्णु, भरतऋषि, कश्यपमुनि, हनुमान, अंगद, नारद, तुम्बुरुगन्धर्व, दुर्गा, पार्वती, मतंङ्ग, यास्क, शार्दूल और कोहल ऋषि, इन ऋषियों ने ताल की रचना की है।

1. Shiv (Sadashiv)
2. Vishnu (Hari manifested as Shri Ram and Shri Krishna)
3. Brahma
4. Sage Bharat Rishi
5. Sage Kashyap Muni

Lord Shri Ram's Eternal Devotee Pavanputra Shri Hanumanji

6.   Pavanputra Shri Hanumanji
7.   Sage Angad
8.   Sage Narad Muni
9.   Gandharva Tumbru
10. Goddess Durga
11. Goddess Parvati
12. Sage Matang
13. Sage Yask
14. Sage Shardul
15. Sage Kohal

"Taal" - The Word and its root meaning from Sanskrit.

तालस्तलप्रतिष्ठायामिति धातोर्हृज्ञि स्मृतः।
गीतं वाद्यं तथा नृत्यं यतस्ताले प्रतिष्ठितम्।।२।।

भावार्थ : 'ताल' शब्द व्याकरण के अनुसार तल् धातु से बना है जिसका कि अर्थ 'किसी चीज़ में किसी का रहना' होता है। तल् धातु से घञ् प्रत्यय लगाकर 'ताल' शब्द बना है, क्योंकि इस ताल में गाना, बजाना और नाचना तीनों रहते हैं। इसलिए ताल इन सभी में प्रधान है।

The word 'Taal' in Sanskrit grammar originates from the root sound 'Tal' which means a thing that resides inside or on another thing. Another word that is common to explore this meaning of "Tal' is the word 'Bhutal' in Sanskrit referring to a "Tal" place for all beings residing on the 'Bhu' Earth. When the sound "aa" is added to this root sound 'Tal', the main word "Taal' is formed in Sanskrit per its grammatical meters.

The inner sense of this word "Taal' can be understood better by us when we explore it from the perspective of the fact that "Inside" this Rhythmic Taal is where the acts of Singing, Dancing and Playing of other non-Rhythmic instruments "Reside". **Hence, Taal is considered as the foundation of Indian Musical Science and Music is pleasurable for the most common man only when appreciated in the constraints of residing within these boundaries of Taal.** Without residing in the parameters of a Taal, the Vocal music, Instrumental Music or Dance would never have the creative crest and wave formations that are essential in the Universe for any living creation. Taal provides this means of start and end and infinite possibilities and variations within itself for the Vocal Musical, Instrumental Musical or Dance exponent.

Taal – Its Bhavatmak (emotional sensibility) definition from another Indian ancient Musical volume.

तकारः शंकरः प्रोक्तो लकारः पार्वती स्मृतः।
शिवशक्तिसमायोगात् ताल इत्यभिधीयते।। ३।।
भावार्थ : 'त' यह शंकर का रूप है। 'ल' यह पार्वती का रूप है, इस प्रकार
महादेव और पार्वती के मिलने से यह ताल कहा जाता है।

The sound "T' is manifested as representing Lord Shiva and
the sound "L represents the form of Parvati, Lord Shiva's wife.
The union of the Lord (Male and Female forms of his
representation) forms the main rhythmic science of Indian
Taal.

Lord Kashi Vishwanath as "Natraj"
King of all Laya in "Lasya" Dance Mudra

तौर्यत्रिकं तु मत्तेभस्तालं तस्याङ्कुशं विदु:।
न्यूनाधिकप्रमाणं तत् प्रमाणं क्रियते यत:।।६।।
भावार्थ : जिस प्रकार मतवाले हाथी को अंकुश वश में रखता है, इसी प्रकार
गाना बजाना और नाचना इन तीन मतवाले हाथियों को ताल रूपी अंकुश
न्यूनाधिक (कम ज़्यादा) समय बतलाता हुआ ठीक मार्ग पर ले जाता है।

A poetic definition of Taal in the Indian texts is analogized
with the word "Ankush", which means controlling an
intoxicated Elephant. In the same way, "Taal" controls the
powerful elephants of Gaan (Vocal Singing), Vadya
(Instrument playing) and Nritya (Dance). Increasing or
decreasing the control with its tempo, Taal (rhythm) brings the
above three elements in Indian Classical music back on their
main path of progression.

कालिय मद मर्दन कियो, कृष्ण कुंवर नन्द लाल।
नृत्य काल पद घात से, प्रकट भये सब ताल।।

Another poetic definition with a devotional element of Taal
comes in the form of Lord Krishna and one of his episodes of
taming a poisonous snake Kali in river Yamuna. The poison
of the snake was tainting the purity of the river Yamuna
beloved of Shri Krishna. It also symbolizes the impurities in
our lives in this present age of Kaliyuga. Now when Lord Shri
Krishna tames this Kali Naag, he dances on his head. The
strikes of Lord's feet on his head in the form of his dance
movements produce the divine sounds that created all the
Taals according to this devotee's "Bhavatmak" devotional
feelings.

The "Kali Naag" and Lord Shri Krishna dancing on his Head and thus creating all Taals. "Naag Nathaiya" Krishna Leela occurring at Tulsi Ghat in Benaras since time of Saint Tulsidasji.

Kaal, Maarg, and Kriya

Kaal

अथ काल:

हस्तद्वयस्य संयोगे वियोगे चापि सर्वदा।

व्याप्तिमान् य: समादिष्ट: स कालस्ताल संज्ञक:।।९।।

भावार्थ : दोनों हाथों को मिलाने या अलहदा करने में काल सर्वदा विद्यमान रहता है, इसी काल का नाम ताल रखा गया है।

The time taken in the act of clapping or bringing two palms together in act of silent or actual clapping as well as separating the palms henceforth, is the act of defining the "Kaal". The time in which this action occurs is one form of physical manifestation of the concept of Taal.

When the above act of bringing the palms together or separating them in specific time intervals named Anu, Drut, etc. (These terms we shall develop and define later) is conducted in conjunction with one of the three other elements of Indian music viz. Singing, Instrument playing or Dancing, then this ACT of creating the "Kaal" itself is called Taal.

कालस्ताल इति भोक्तो त्र्यराबादिक्रिययामित:।

गीतादे: संमिति कुर्वन् मार्गी देशीति स द्विधा।।१०।।

भावार्थ : गायन के साथ अणु इत्यादि परिमित काल में इसकी (ताल की) क्रिया होने की वजह से यह काल ही ताल कहा जाता है। जो मार्गीय और देशीय भेद से दो प्रकार का है। नोट - जिस काल के अनुसार गन्धर्वादि गाते हैं उस काल को मार्गीय काल कहते हैं और जिसके अनुसार मनुष्य गाते हैं उसे देशीय काल कहते हैं।

Such Kaal is grouped in two major categories as per Indian texts, "Margiya" and "Deshiya". The "Kaal" according to which the semi-divine beings or celestial beings sing and perform music is called "Margiya" and the Kaal according to which humans manifested perfume music is called "Deshiya" Kaal. Hence, all the description that applies to us, the readers and author, in the text refers to the "Deshiya" Kaal (Taal).

A "Kshan" which in English practical sense means the minutest physical measure of observable time, is very difficult to express and define in the act of "Kaal". Thus, the larger time intervals that are easier to create using Kaal are used in the normal form of music creation. These other intervals of time will be explored soon at the appropriate sequence of building our scientific knowledge of Taal. The Kaal intervals that can be used practically are described below.

Developing Kaal units

उपर्युपरि विन्यस्य पद्मपत्रशतं सकृत्।
सूचीकृतस्तत्संवेधे यः काल : स क्षणः स्मृतः।।१२।।

भावार्थ : एक के ऊपर एक इस प्रकार से १०० कमल पत्र रखकर सुई के अग्र भाग से छेदन करने में जो समय लगता है वह क्षण कहलाता है।

लवः क्षणैरष्टभिः स्यात् काष्ठा स्यादष्टभिर्लवैः।
अष्टकाष्ठानिमेष: स्यान्निमेषैरष्टभिः कला ।।१३।।

भावार्थ : आठ छण का एक लव होता है। आठ लव की एक काष्ठा होती है। आठ काष्ठा का एक निमेष होता है और आठ निमेष की एक कला होती है।

कला द्वयाच्चतुर्भागश्चतुर्भाग द्वयादणुः।
अणुद्वयाद् द्रुतः प्रोक्तस्तद्द्वयेन लघुर्भवेत्।।१४।।

भावार्थ : दो कला का एक चतुर्भाग और दो चतुर्भाग का एक अणु, दो अणु का एक द्रुत और दो द्रुत का एक लघु होता है।

लघुद्वयाद् गुरुः प्रोक्तस्त्रिलघुः प्लुत उच्यते।
एवं काल गतिः प्रोक्तः तालज्ञैः पूर्वसूरिभिः।।१५।।

भावार्थ : दो लघु का एक गुरु और तीन लघु का एक प्लुत होता है। इस प्रकार ताल के जानने वाले विद्वानों ने ताल का वर्णन किया है।

इति कालः।

The question now arises whether all of these concepts are just theoretical or do they have actual physical practical base in creating time intervals of Kaal (Taal)? And for the inquisitive mind, the ancient texts have precisely defined the smallest definition of the practical unit of time that one can start with for creating structures of Rhythmic time. This first starting point is what we call as the initial measure of smallest unit of measurable time in the Indian musical context. This is referred to as "Kshan", which in practical scientific context for music is defined as follows. Take 100 rose petals and stack them one by one on top of each other than pierce the stack with the pointed part of the sewing needle. Time taken to pierce this stack is the smallest unit in Indian musical science that is of use and it is called "Kshan".

a.	EIGHT "Kshan"	= 1 "Lav"
b.	EIGHT "Lav"	= 1 "Kashta"
c.	EIGHT "Kashta"	= 1 "Nimesh"
d.	EIGHT "Nimesh"	= 1"Kala" (this is Kala
		which is the subunit of "Kaal" above used to define Maarg in next section)
e.	TWO "Kala"	= 1 "Chaturbhag"
f.	TWO "Chaturbhag"	= 1 "Anu"
g.	*TWO "Anu"*	*= 1 "Drut"*
h.	*TWO "Drut"*	*= 1 "Laghu"*
i.	*TWO "Laghu"*	*= 1 "Guru"*
j.	*THREE "Laghu"*	*= 1 "Plut"*

In the above manner, the creators of Indian classical music have defined in the smallest measurement progressions the units of Taal. The above names in Sanskrit are the actual definitions of components of Taal as prescribed by all of the Fifteen Founding Music Acharyas described earlier. Since the creation of the universe, these define the units of composition of a Taal structure in Indian musical sense. The

Kaal units listed in **_BOLD ITALICS_** above are of most use in common practice of music in practical terms.

Maarg

अथ मार्गा:
मार्गा: स्युस्तत्र चत्वारो ध्रुवश्चित्रश्च वार्त्तिक:।

दक्षिणश्चेति तत्र स्याद् ध्रुवके मात्रिका कला।।१६।।
शेषेषु द्वेचतस्रोऽष्टौ क्रमान्मात्रा: कला भवेत् ।।

भावार्थ : ध्रुव, चित्र, वार्त्तिक और दक्षिण इन भेदों से मार्ग चार प्रकार का है। इनमें से ध्रुव नामक मार्ग में एक कला होती है और शेष मार्गक्रम से दो, चार और आठ कलाओं से बनते हैं।

यहाँ पर संगीताचार्यों की कुछ मत भिन्नता दिखाई देती है क्योंकि चूड़ामणि में ६ और मणिदर्पण में १२ मार्ग कहे हैं।
(इति मार्गा:।)

"Maarg" in Sanskrit literally means path. And after defining the Kaal definitions above, some combinations of those Kaals make up different paths or "Maarg" of Taal. In other words, Maarg defines the manner in which after the first beat of Taal how the following beats of Taal progress with space of time in between them, with specific measures and tempos in-between the first beat and the last beat. This flow of Taal and its progress in its single base structure is called its Maarg. There are four Maargs described in ancient texts.

1. Dhruv   = 1 Kala (refer to bullet D above)
2. Chitra   = 2 Kala
3. Vaartik   = 4 Kala
4. Dakshin   = 8 Kala

There is some disagreement in ancient texts in terms of how many paths of Taal creation exist and most predominant are the four listed above. However in one text called Sangeeta Chudamani, the "Maarg" are listed as six and in another text Manidarpan, 12 "Maarg' are listed. We stick with four above as widely used definition.

Kriya

अथ क्रिया

नि:शब्दा शब्दयुक्ता च क्रिया तु द्विविधा मता।

नि:शब्दा तु कला प्रोक्ता चतुर्धा सा प्रकीर्तिता।।१७।।

भावार्थ : 'नि:' शब्द और 'स' शब्द भेद से क्रिया दो तरह की होती है। नि: शब्द क्रिया कला नाम से भी पुकारी जाती है, जो चार प्रकार की है।

आवापश्चाथ निष्कामो विक्षेपश्च प्रवेशक:

नि:शब्देति चतुर्धोक्ता-

भावार्थ : आवाप, निष्काम, विक्षेप और प्रवेशक भेदों से चार प्रकार की कही गयी है।

सशब्दापि चतुर्विधा।।१७।।

ध्रुव: शंपा तथा ताल: सन्निपात इतीरिता।

भावार्थ : 'स' शब्द क्रिया भी ध्रुव शंपा ताल तथा सन्निपात इन भेदों से चार तरह की है।

This is a very important and third Prana of Taal definition. Basically "Kriya" in Sanskrit means Action. The Kriya can be of two types in the Science of Taal. "Nishabd" (Silent action of indicating a Taal) or "Shashabd" (Spoken Action with sound to indicate and signify a Taal.

A. Nishabd Kriya (Silent) - is further broken down into four different categories of acts of silently indicating Taal formation.

पातः कला च सा ज्ञेया तासां लक्ष्माभिधीयते।।१८।।
आवापस्तत्र हस्तस्योत्तानस्यागुंलिकुंचनम् ।
निष्कामोऽधस्तलस्य स्यादंगुलिनां प्रसारणम्।।१९।।

भावार्थ : 'स' शब्द क्रिया को पात तथा कला इन नामों से पुकारते हैं। अब इन सब भेदों के लक्षणों को कहते हैं।

आवाप - उठे हुए हाथ की अंगुलियों के बटोरने की क्रिया को कहते हैं।

निष्काम - हाथ को नीचे कर अंगुलियों के फैलाने की क्रिया को कहते हैं।

1. Aavap – This is defined as the act of **bringing together** in a rhythmic manner the spread-out fingers of a **raised palm of a hand** silently signifying the formation of some units of Taal.

2. Nishkaam – The act of keeping your **palm facing down and spreading the fingers** to signify rhythm.

क्षेपो दक्षिणपार्श्वस्योत्तानस्य प्रसुतांगुलेः।
विक्षेपोऽधस्तलस्यास्य प्रवेशोऽङ्गुलिधूननम्।।२०।।

विक्षेप - फैले हुए दाहिने हाथ की उठी हुई अंगुलियों के गिराने को कहते हैं।

प्रवेशक - उन्हीं फैली हुई अंगुलियों के नीचे के हिस्सों को कँपाने के प्रयोग को कहते हैं।

3. Vikshep – The act of **dropping the fingers of the right-hand palm** when the palm and fingers are **starting from the raised original position** to signify Taal.

4. Praveshak – This variation signifies the **palm facing down but fingers are shaken here** signifying rhythm formations.

B.   Shashabd Kriya (With sound) – is also further broken down into four different categories.

95

ध्रुवोहस्तस्य पातः स्याच्छोटिका शब्द पूर्वकः।
शंपा दक्षिण हस्तस्य तालो वाम करस्य तु।
उभयोः सन्निपातस्यात्तासां मार्गवशान्मितिः।।२५।।

भावार्थ - ध्रुव - हाथ के गिराने को कहते हैं।

शंपा - दाहिने हाथ की शब्द पूर्वक चुटकी बजाने को कहते हैं।

ताल - बाँये हाथ की शब्द पूर्ण चुटकी बजाने को कहते हैं।

सन्निपात - दोनों हाथों की एक साथ चुटकी बजाने को कहते हैं।

नोट - वर्तमान समय में चुटकी बजाने के स्थान पर एक हाथ पर दूसरा हाथ गिराकर ताल का कार्य लेते हैं।

(इति क्रिया) ।

1. Dhruva – This is the *act of dropping the hands and palms downwards* to strike either thigh in a lotus position posture or clapping with one hand used as a base for the production of a sound formation.
2. Shampa – *Right-hand fingers are used here to produce "snap"* and signify rhythm with sound here.
3. Taal – Left-hand *fingers are used here to produce "snap"*
4. Sannipat – *Both hands are used to "snap" the fingers with the thumb* to count and produce the rhythmic structure.

P.S. In present times, practitioners demonstrate the Kriya by putting one palm over the other and counting. Also, this would be an ideal point to revisit the chapter on the latest neuroscience findings described earlier in the book on Motor skills and limbic response. Now as we go deeper into the science of Taal and its Prana we can connect these ancient ideas of the limbic acts "Kriya" with what our modern science is observing and documenting in relation to Rhythm.

Ang, Graha, Jati

In this Section two of discussing the next group of the Ten Prana of Taal, we now explore in the following order the next set of three Prana of Ang, Graha, and Jati.

Ang

अथ अङ्गानि'

अब अङ्गों का वर्णन किया जाता है

श्लोक - अनुद्रुतो द्रुतश्चाथ दविरामो लघुस्तथा।

लविरामो गुरुश्चैव प्लुतश्चेति यथा क्रमम्।।

सप्ताङ्गानीह तालेषु ज्ञातव्यानि सदा बुधैः।

अण्वाद्याद्यक्षरैर्ज़ेयं सन्ति संज्ञान्तराण्यपि।।

अनुद्रुत, द्रुत, द्रुतविराम, लघु, लघुविराम गुरु और प्लुत यह ताल के ७ अङ्ग हैं, जिनका जानना विद्वानों के लिए आवश्यक है, अणु आदि अक्षरों के अतिरिक्त इनके दूसरे नाम भी लिखे जाते हैं।

The word "Ang" in English is literally defined as a limb of a body. "Taal" in Indian music science has Seven Ang. *The Seven Ang are defined as described here along with their associated Deities from the Indian spiritual realm.* Each name has a specific Sanskrit meaning whose precise description and definition would deviate us too much from our main subject of discussing the nature of Taal and Indian Rhythmic structure definition. Hence, for the inquisitive student, it is suggested to refer to a Sanskrit-English dictionary to decipher some precise meanings of the names mentioned in our book here.

अंगों के देवताओं के नाम--

श्लोक - अणों चन्द्रो द्रुते शम्भुर्द्विरामें षडाननः।

लघौ देवी लविरामे जीवो गौरीपतिर्गुरो।।

प्लुतेत्रयोविरिंच्याद्याः देवता मुनिभिः स्मृता।

भावार्थ - अणु के देवता चन्द्रमा, द्रुत के देवता शम्भु, द्विराम (द्रुतविराम) के देवता षडानन (कार्तिकेय) लघु के देवी, लविराम (लघुविराम) के जीव, गुरु के देवता गौरीपति (महादेव), प्लुत के देवता ब्रह्मा, विष्णु और महेश तीनों हैं।

1. Anudrut   (Chandra – Moon)
2. Drut   (Shambhu – Lord Shiva)
3. Drutviram   (Shadanan – Lord Kartikeya elder
                     Son of Shiva)
4. Laghu   (Devi – All Goddesses)
5. Laghuviram   (Jeeva – Life)
6. Guru, and   (Gauripati – Lord Shiva)
7. Plut   (Trinity–Brahma, Vishnu, Mahesh)

Each of the above Seven are also referred in many ancient texts with other synonymous names, which we shall describe as follows;

अनुद्रुत के दूसरे नाम-

श्लोक - अनुद्रुतोऽणुः करजमर्धचन्द्रोंऽकुशं धनुः।

भावार्थ - अणु, करज, अर्धचन्द्र, अंकुश, धनु।

द्रुत के दूसरे नाम--

श्लोक- अर्धमात्रं व्यञ्जनं खं बिन्दुश्च वलयं द्रुते।

भावार्थ - अर्धमात्र, व्यञ्जन, ख (आकाश = शून्य) बिन्दु, वलय (कड़े के समान)।

लघु के दूसरे नाम--

श्लोक - व्यापकः सरलो ह्रस्वः शरो दण्डो लघु स्मृतः।

भावार्थ - व्यापक, सरल, ह्रस्व, शर, दण्ड।

गुरु के दूसरे नाम--

श्लोक - दीर्घो वक्रो द्विमात्रो गोजीवः पूज्यो भवेद्गुरुः।

भावार्थ - दीर्घ, वक्र, द्विमात्र, गोजीव (गौ के मूत्र करते समय मूत्र आकार में टेढ़ा होता हुआ भूमि पर आता है, उसी प्रकार गुरु भी टेढ़े आकार का होता है।) पूज्य।

Anudrut is also known as Anu, Karaj, Ardhchandra, Ankush, and Dhanu.

Drut is also known as Ardhmatra, Vyanjan, Kha (Aakash = 0) Bindu, Valay.

Laghu is also known as Vyapak, Saral, Hrsv, Shar, and Dand.

Guru is also known as Deergha, Vakra, Dwimatra, Gojiv, and Poojya.

प्लुत के दूसरे नाम--

श्लोक - त्र्यङ्गत्रिमात्रको दीप्तः शृङ्गी सामोद्भवः प्लुतः।

भावार्थ - त्र्यङ्ग, त्रिमात्रक, दीप्त, शृङ्गी, सामोद्भव।

Plut is also known as Tryang, Trimatrak, Deepta, Shringi, and Saamodhav.

Now we proceed to define the quantifiable units of measuring and using these Ang as building blocks in the definition of the Rhythmic structure of Indian Taal. *Laghu is used as the reference unit of measurement against which all others are defined*.

Concept of Matra

श्लोक- एक मात्रो लघु: प्रोक्तो द्विमात्रो गुरुरुच्यते।
त्रिमात्रा: प्लुत उद्दिष्टो मात्रार्थं तु द्रुतो मत:।।
अनुद्रुतश्चतुर्थांशो मात्रायाः इति संस्मृतम् ।
पञ्चलघ्वक्षरोच्चार मिता मात्रास्ति मार्गिके।।
चतुर्भिरक्षरैर्दैशी षड्भिरप्यक्षरैः क्वचित् ।
अनया मात्रया ज्ञेया लघुगुर्वादि कल्पना ।।

भावार्थ - एक मात्रा का लघु, दो मात्रा का गुरू, ३ मात्रा का प्लुत और आधी मात्रा का द्रुत और चौथाई मात्रा का अनुद्रुत माना गया है। पाँच लघु अक्षरों के उच्चारण करने में जितना समय लगता है मार्गीय में वह समय १ मात्रा का होता है। दैशी में चार लघु अक्षरों के उच्चारण करने में अथवा ६ लघु अक्षरों के उच्चारण करने में जितना समय लगता है वह १ मात्रा कहलाता है। किन्तु ६ अक्षरों वाली मात्रा का उच्चारण उन्हीं चार अक्षरों के समय में होना चाहिये। किन्हीं २ तालज्ञों ने जितनी देर में स्वस्थ पुरुष की नाड़ी दो बार चले उतने समय को एक मात्रा माना है।

1.	Anudrut	= ¼ Laghu
2.	Drut	= ½ Laghu
3.	Drutviram	= ¾ Laghu
4.	***Laghu***	*=1 Matra (base reference unit of measurement)*
5.	Laghuviram	= 1.5 Laghu
6.	Guru, and	= 2 Laghu
7.	Plut	= 3 Laghu

The question that arises now is "what is this Matra if it is considered as the base unit?" According to many ancient musical texts and scholars, several definitions of 1 Matra exist.

1 Matra (Maargiya Kaal definition) = This is the time taken to pronounce and speak Five Laghu words per Sanskrit Grammatical science as per the "Maargiya" Kala defined in the last chapter. This is the measurement of time used by Celestial beings (Gandharva) for creating their music.

1 Matra (Deshiya Kaal Definition) = Time taken to pronounce and speak Four or Six Laghu words per Sanskrit Grammatical science as per the "Deshiya" Kala (Human music practice) defined in the last chapter. However even if Six Laghu words are spoken, their time taken must equal Four words interval. This is the measurement of time used by Human beings in creating their music and applies to the reader and the author.

According to some practical definitions outside of theory, some old musicians and masters of Rhythm consider;

1 Matra = Time of Two Heartbeats of a healthy person.

The student reading this volume must understand that a truly rounded modern Indian Rhythmic player must be aware of these deep scientific and systematic concepts of the evolution of the Taal that they are physically playing. Emphasizing only on the physical form and ignoring the roots of the process of why and what is being played results in just "Exercise" of playing an instrument. However the true seeker goes beyond playing, their goal is to "Feel" the instrument so that it can speak to him or her and this is not possible if one's only focus is on physical exercise. **This commitment of Physical Practice is not a bad thing in the ordinary sense, but it must be coupled with the appreciation of the deeper scientific meaning and spiritual sensibilities behind the derivation of our understanding of what we now call "Taal' what is being played.**

Most such modern players who are unaware of these deeper theories of Taal, only refer to the concept of Matra not always knowing precisely what it means and how our sages derived at this term Matra. So, when one hears the term Matra in common day-to-day usage in Indian "Taal" definition, we need to remember most importantly how the Matra defines all the main Ang (limbs) of the Taal.

Having discussed the theoretical aspects of Seven "Ang" (limbs) of Taal, their respected spiritual deities and their basic measurements with respect to the base unit of 1 Matra = 1 Laghu, we now turn our attention to more practical meaning of these Ang.

The physical actionable definition of the main Ang is as described. Note that Laghuviram and Drutviram are extremely difficult to produce physically on a rhythmic instrument and for all practical purposes, the discussion revolves around the five Ang's besides Laghuviram and Drutviram. And even within the five, practically speaking emphasis is on Drut, Laghu and Guru as three main physical Ang to start building a Taal composition.

श्लोक - अत्यल्पघातेऽणुर्ज्ञेय: सूक्ष्मघाते द्रुतो मत:।
पूर्णघाते लघु: प्रोक्तो घातात्क्षेपे गुरुर्मत:।।
घातात्करभ्रमो यत्र प्लुतो ज्ञेयो विचक्षणै:।

भावार्थ - बहुत थोड़े घात से अणु और सूक्ष्म घात से द्रुत, पूर्ण घात से लघु और घातक्षेप से (फेंकने से) गुरु और जहाँ घात से करभ्रम हो वहाँ प्लुत जानना चाहिए।

1. Drut = Brief faint strike of rhythmic instrument
2. Anudrut = Extremely brief strike
3. Laghu = Full strike of instrument
4. Guru = Elongated strike
5. Plut = Elongated strike with a feeling of triple long compared to Laghu strike.

अङ्गों के स्वरूप -

श्लोक - अर्धचन्द्रस्वरूपोऽणुः द्रुतः स्याद्वलयाकृतिः।
द्रुते विरामेऽणुर्ज्ञेयो लघुस्तु सरलाकृतिः ।।
लघौ द्रुतो विरामस्तु गुरुर्वक्रः प्लुतस्त्रिकम् ।
गुरौ रेखेति चिन्हानि लिप्यामुक्तानि सूरिभिः ।।

भावार्थ - अणु का स्वरूप आधे चन्द्रमा के समान (˘) और द्रुत का कुण्डल के समान (०), दविराम का स्वरूप द्रुत के ऊपर एक अणु (०˘), लघु का स्वरूप सरल आकृति का (।) और लविराम का स्वरूप लघु के ऊपर एक द्रुत (१) और गुरु का आकार टेढ़ा (S) और प्लुत तीन (३) के आकार का होता है।

अङ्गों को उच्चारण करने वाले पक्षियों के नाम -

श्लोक - तित्तिरष्टकश्चैव वक चातक कोकिलाः।
वायसः कुक्कुटश्चैव क्रमादुच्चारयन्त्यमून्।।

भावार्थ - अणु को तीतर, द्रुत को चटक = गौरैया, दविराम को वक = वगुला, लघु को चातक = पपीहा, लविराम को कोयल और गुरु को वायस = कौआ और प्लुत को कुक्कुट = मुर्गा उच्चारण करता है।

Thus far we have covered the Ang, their meanings and their practical measures in physically playing the percussion instrument. Now we discuss how these are represented in written instructions and creative compositions. Each of the Ang is represented in written formations of Taal with specific symbols that almost look like Egyptian Hieroglyphics. These symbols are important for the student wishing to master the science of Taal. The symbols along with Sanskrit/ Hindi names of the terms that we have defined for each Ang are given in the pictured *Ang Table* here. The last column on the right is the actual weight (or length) of the sound in the action of playing that particular rhythmic beat. Second column from the right next to the Weight column is the actual Symbolic representation of that Ang.

7 Ang of Taal and their corresponding written Symbols.

अङ्ग निदर्शक सारिणी

संख्या	नाम	देवता	मात्रा	उच्चारक पक्षियों के नाम	चिन्ह	वज़न
१	अणुद्रुत	चन्द्रमा	चौथाई	तित्तिर	⌣	I)
२	द्रुत	शिव	आधी	चटक	o	II)
३	द्रुतविराम	कार्तिकेय	पौन	बगुला	ỏ	III)
४	लघु	देवी	एक	पपीहा	I	१)
५	लघुविराम	बृहस्पति	डेढ़	कोयल	ा	१II)
६	गुरु	गौरीपति	दो	कौआ	S	२)
७	प्लुत	ब्रह्मा, विष्णु महेश	तीन	मुर्गा	S̄	३)
	काकपद				+	

नोट - यह काकपद केवल सम के निशान में लिया जाता है किन्तु मार्गीय तान में यह प्रस्तार के रूप में लिया जाता है।

Graha (Sama, Atit, Anagat)

अथ ग्रहा:
अङ्गोंके पश्चात् ग्रहों के लक्षण लिखे जाते हैं।
श्लोक- - समोऽतीतोऽनागतश्च विषमश्च चतुर्विधः।
ग्रहास्तालेषु विज्ञेयाः सूक्ष्मदृष्ट्या विचक्षणै: ।।
भावार्थ - सम, अतीत, अनागत और विषम ये चार प्रकार के तालों के ग्रह संगीत
पारिजात ने माने हैं, किन्तु संगीत रत्नाकर ने विषम ग्रह छोड़कर केवल तीन ही ग्रह माने हैं।

We have discussed four Prana of Taal so far namely Kaal, Maarg, Kriya and Ang. **Now we move forward to understand the next Prana of Taal called Graha. Graha (also spelled in English transliteration as Griha many times) in Sanskrit: ग्रह means planet, seizing, laying hold of or holding.** There are four main Graha of Taal as identified in the ancient musical text Sangeet Parijat. Another text, Sangeet Ratnakar, refers to just three Grahas leaving out one of the four mentioned by the former text.

1. Sama
2. Atit   (also called AvPani)
3. Anagat (also called UpariPani)
4. Visham (not mentioned in Sangeet Ratnakar)

Let us further understand what each one of these Graha means in the context of the rhythmic structure of a Taal.

Sama Graha

When the *START* of the vocal singing, instrument or dance coincides precisely with the start of the Taal, it is called Sama Graha. This is the Graha most appreciated by a common music listener who maybe is not as well versed in the art and science of Music. The act of each form of Music resonating in its starting cycle precisely with the start of the beat cycle is the "Sama". When most musicians and listeners swing with joy and arriving at this moment of unison, their souls are naturally appreciating the beauty of this moment. **Sama has a deep philosophical meaning which can have its own chapter if we expound on its philosophical meanings in Life. This might deviate us too much from the present discussion but suffice it to say that the seeker who strives to perfect the ideals of Sama in Music carries it forward into a deeper understanding of Nature itself surrounding us and ultimately the creator himself.**

श्लोक - गीतादौ विहिते यत्र ताल वृत्ति र्विधीयते ।
अतीताख्यो ग्रहो ज्ञेय: सोऽवपाणिरिति स्मृत:।।
भावार्थ - गीतादि पहले शुरू हो और उसके पश्चात् ताल प्रारम्भ हो इस क्रिया को अतीत ग्रह कहते हैं, इसका दूसरा नाम अवपाणि भी है।
नोट - अवपाणि का अर्थ है अवगत = जान लिया गया है, पाणि = ताल।

Atit Graha

Musical sections where the Rhythmic strike and beat portion occurs slightly after the start of the other Musical non-rhythmic components is called Atit Graha. This is also known as AvPani (Av refers to term Avgat, meaning after the fact and Pani means the music produced by striking the rhythmic instrument with one's palms such as Mridang or Tabla, etc.)

Anagat Graha

श्लोक - पूर्वं ताल प्रवृत्ति: स्यात् पश्चात् गीतादिरुच्यते।
अनागत: सविज्ञेयो स एवोपरिपाणिक:।।

भावार्थ - पहले ताल तत्पश्चात् गीतादि प्रारम्भ हो तो इस क्रिया को अनागत ग्रह कहते हैं, इसका दूसरा नाम उपरिपाणि भी है। (उपरि = पहले, पाणि = ताल अर्थात् पहले ताल की प्रवृत्ति हो)

Musical sections where the Rhythmic strike and beat portion occurs slightly before the start of the other Musical non-rhythmic components is called Anagat Graha. This is also known as UpariPani (Upari in Hindi means first and Pani means the music produced by striking the rhythmic instrument with one's palms such as Mridang or Tabla, etc.)

ग्रहों के दूसरे नाम-

श्लोक -   तालो वितालोऽनुताल: प्रतितालश्चतुर्विध:।
समग्रहो भवेत्तालो वितालोऽतातक: स्मृत:।।
अनागतोऽनुतालस्यात् विषम: प्रतितालक:।

भावार्थ - समग्रह का दूसरा नाम ताल, अतीत का विताल, अनागत का अनुताल और विषम का दूसरा नाम प्रतिताल है।

These Grahas are also called by other names which are common day to day referential names signifying their nature and understanding described in the verse above as follows;

1. **Sama = Taal** (Sama is the actual normal nature of Taal and rhythmic progression in unison with other non-rhythmic structures. Hence it is equated with Taal itself)
2. Atit = Vitaal
3. Anagat = AnuTaal
4. Visham = PratiTaal is also known as "Out of Synch" in English phraseology. (Certainly not used in day to day usage and the term Prati here means opposite of Taal).

One can naturally suggest here that the purpose of our Rhythm and Melody is NOT to be in Visham Graha and hence it is to be avoided at all costs. *And for a true student, this journey from Visham to Sama is the essence of the fruit of the Naad Yoga.*

Jati

अथ जातयः
श्लोक - चतुरस्रस्तथा त्र्यस्रः खण्डो मिश्रस्तथैव च।
संकीर्ण इति विज्ञेयाः क्रमशो बुधैः ।।
भावार्थ - चतुरस्र, त्र्यस्र, खण्ड और मिश्र तथा संकीर्ण यह तालों की पाँच जातियाँ
मानी गई हैं।
श्लोक - तथैव मात्रा विज्ञेयाः क्रमशस्तालवेदिभिः।
द्विगुणं द्विगुणाद्यत्र लक्ष्य दृष्ट्यासु संमतम्।।
भावार्थ - उसी प्रकार क्रम से जातियों के अनुसार तालों की जाति की मात्रा जानना चाहिये।

Now we move on to Jati of Taal. The word Jati essentially signifies types of Taal. So by referencing the Jati, we are referring to what *Type* of Taal it is when we deal with different rhythmic structures. **There are Five Jatis of Taals.** Understanding the differences of these Jatis is needed first before properly playing or composing the Taal structures. These Five Jatis are as named here;

1. Chaturastra   = 4 Matra (Chatur means Four)
2. Triyastra   = 3 Matra (Tri means Three)
3. Khand   = 5 Matra
4. Mishra   = 7 Matra
5. Sankirna   = 9 Matra

श्लोक - चतुर्वर्णैस्त्रिभिर्वैः पंचवर्णैस्तथैव च।
सप्तवर्णैश्च नवभिः जातयः क्रमशः स्मृताः।।
भावार्थ - चतुरस्र की ४ मात्रा, त्र्यस्र की ३ मात्रा, खण्ड की ५ मात्रा, मिश्र की ७
और संकीर्ण की ९ मात्रा हैं।

What it means is that each Jati of Taal phrase has a specific number of Matras. One can combine in the creation of rhythmic structures phrases (or Bols) of different Jatis and create unique patterns suitable for that occasion. Also, if a Taal has odd or even beats, the odd or even Jati phrases are used accordingly.

जातियों के वर्ण -

श्लोक - चतुरस्रो ब्राह्मणः स्यात् त्र्यस्रः क्षत्रिय एव च।
खण्डो वैश्यस्तथा मिश्रः शूद्र इत्यभिधीयते।।
संकीर्णजातिः संकीर्ण इत्येताः पंचजातयः।

भावार्थ - चतुरस्र जाति का वर्ण ब्राह्मण, त्र्यस्र का क्षत्रिय, खण्ड का वैश्य तथा मिश्र का शूद्र और संकीर्ण का वर्ण संकर = अन्त्यज है।

नोट - जिस ताल की जाति जाननी हो उस ताल की मात्रा का छोटे से छोटा रूप बनाना चाहिए, किन्तु ताल का छोटे से छोटा रूप वहाँ तक बनाना चाहिए, जहाँ तक कि वह कहा जा सके। उस छोटे रूप की मात्राओं की तुलना इन भिन्न-भिन्न जातियों की मात्राओं के साथ करके उस ताल की जाति का निर्णय करना चाहिए। जैसे त्रिताल १६ मात्राओं का है, इसका छोटे से छोटा रूप ४ मात्राओं का होता है और चतुरस्र जाति की मात्रा ४ हैं, अतः यह ताल चतुरस्र जाति का हुआ, इस प्रकार चौताल त्र्यस्र जाति का है, क्योंकि इसका छोटे से छोटा रूप ३ मात्राओं का होता है, दादरा और खेमटा भी त्र्यस्र जाति के हैं। झपताल और शूल ताल का छोटे से छोटा रूप ५ मात्राओं का होता है, अतः ये दोनों ताल खण्ड जाति के हैं।

यत, धमार, ब्रह्मताल, आड़ा, चौताल, रूपक और तेवरा ये सब ताल मिश्र जाति के हैं। इनका छोटे से छोटा रूप ७ मात्रा का होता है। मत्तताल और लक्ष्मीताल का छोटे से छोटा रूप ९ मात्रा का हो सकता है, अतः ये दोनों ताल संकीर्ण जाति में रखे जाएँगे।

If one wishes to identify the Jati of a Taal or create a Taal of a specific Jati, the process is described here.

    A. First, identify the smallest possible phrase of the Taal structure as can be practically spoken with one's mouth and played as well. This is usually possible when the Taal is played in the fastest tempo making the smallest phrase repetition easier in Drut (Fast tempo speed). So, identify the Drut form of the Taal and then define its smallest phrase.

    B. Then, identify the number of Matras in that smallest phrase of the Rhythmic structure.

C. Compare the identified Matra with the types of Jatis and their Matras and then that should signify what type of Taal is that particular structure.

Let's look at some examples of well-known Taals in the Indian musical system and identify their Jatis.

a. **Tritaal (Teen Taal 16 Beats, 4x4 structure)** = its smallest phrase has 4 Matras (4 beats) when one deconstructs the structure of Tritaal. Now when we compare it to the Jatis, we see that Chaturastra Jati has 4 Matras and hence Tritaal = Chaturastra Jati.

b. **Chautaal (12 beats)** = Triyastra Jati. The smallest phrase in Chautaal has three Matras and hence it is a Triyastra Jati type of Taal. Other Taals of Triyastra Jati are Dadra and Khemta as examples with the smallest phrases comprised of 3 Matras.

c. **Jhaptaal (10 beats)** = Khand Jati. Another Taal with 10 beats is Sool Taal used in Dhrupad music in its final drut format at the concluding portion of Dhrupad performance. Both these Taals have the smallest phrase of 5 Matras and hence are of Khand Jati.

d. **Dhamar (14 beats)** = Mishra Jati. The smallest phrase is 7 Matras. Other Taals of Mishra Jati are Ada Chautaal (14 beats), Tevra (7 beats), Rupak (7 beats), etc.

e. **Matta Taal (21 beats)** = Sankirna Jati. Another similar Taal is Laxmi Taal and both of these have the smallest repeatable portion of the Taal phrase of 9 Matras and hence belong to Sankirna Jati.

Kala, Laya, Yati, Prastar

Kala

कला विवेचन

ध्रुव का, सर्पिणी, कृष्णा, पद्मनी च विसर्जिता।
विक्षिप्तका, पताका च कला स्यात्पतिताष्टमो।।
स शब्दा, तु ध्रुवा, ज्ञेया सर्पिणी वाम गामिनी।
कृष्णा, दक्षिण तो गन्त्रो पद्मनी सा दधो गता।।
विसर्जिता, बहिर्याता विक्षिप्ता कुञ्च नात्मिका।
पताका, तूर्ध्व गमना त्पतिता कर पात नात् ।
ध्रुव पाते प्रयोज्यास्ता नाबा पादौ कदा चनौ।
लघौ तु ध्रुवका ज्ञेया ध्रुवका पतिता गुरौ ।।
ध्रुव का सर्पिणी कृष्णा स्तिस्न एताः प्लुत मताः।
तत्र मार्ग प्रभेदेन कला भेदान्त्रचक्ष्महे।।

इस प्रकार संगीतज्ञ विद्वानों ने कला के ८ भेद कहे हैं। उनके नाम ध्रुवका, सर्पिणी, कृष्णा, पद्मनी, विसर्जिता, विक्षिप्तका, पताका और पतिता ये सब कलायें शाब्दिक कला के अन्तर्गत प्रयोग में लाई जाती हैं। शब्द के साथ ध्रुवका, कला, बाँयी ओर चलने वाली सर्पणी, दाहिनी ओर जाने वाली कृष्णा तथा नीचे को जाने वाली पद्मनी, बाहर की ओर जाने वाली विसर्जिता कला कहलाती है। हाथों को सिकोड़ने से होने वाली कला विक्षिप्ता, ऊपर जाने वाली पताका और हाथ के गिराने से पतिता कला होती है।
सूचना - क्रिया की निपुणता को कला कहते हैं।

इन सब कलाओं का प्रयोग सर्वदा ध्रुव पद अर्थात् शाब्दिक क्रिया में हुआ करता है। निःशाब्दिक क्रिया के आवाप इत्यादि चारों भेदों में शब्द नहीं होता। ध्रुवका कला लघु (।) में ध्रुवका और पतिता गुरू (S) में तथा ध्रुव का सर्पिणी और कृष्णा ये तीनों कलायें प्लुत में S होती हैं अब यहाँ मार्गीय भेदों से कला के भेदों को कहता हूँ।

Kala is the next Prana of Taal. Term Kala in Sanskrit has a much varied multi-layered meaning, but in English translation, Kala is often referred to as the physical prowess or "Art" of excelling in one's chosen form of artistic performance. In terms of Indian Rhythmic structures, Kala refers to the excellence of physical performer in playing that instrument

112

namely Mridang or Tabla, etc. These are played with palms of our hands, both left and right, and hence the description of Kala in Taal signifies many different facets of excellence in playing the instrument with physical techniques that we will now explore.

According to the sages and experts of Indian Music origins, eight types of Kala are identified for the Rhythmic performance.

1. Dhruvka – The act of excellence in producing rhythmic intonations with hand in synchrony with vocal words and singing refers to Dhruvka Kala.
2. Sarpini – Performance of left-handed strikes
3. Krishna – Performance of right-handed strikes
4. Padmini – Downward movements are referred to as Padmini
5. Visarjita – Outward movements or palms or hands stretching out
6. Vikshipta – Movements where palms and hands are closed or narrowed inwards
7. Pataka - Movement Upwards
8. Patita – Movement of falling hands

All the above Kalas are used in "Dhruv Pad" which is essentially a "Shashabd" Kriya as we have discussed in earlier Prana of Taal. In Shashabd Kriya the Dhruvka Kala is Laghu ang, Patita is Guru Ang and both Sarpini and Krishna Kalas are Plut Ang. In "Nishabd" Kriya, there are no words and the act of leaving and starting or migrating from one Kriya to another as part of the above Kala's is called "Aavap".

Mastery over these Kalas of Taal is essential for a truly all-rounded artist par excellence. What we often find is that modern-day Indian Rhythm artists are often focused on all other Kala's except for first one Dhruvka which requires tremendous amounts of practice and as you will see is the essential baseline for measuring the excellence of performer. All other Kalas are measured in reference to Dhruvka as we will see here.

तदर्धार्ध प्रभेदेन शाश्वतः सम्प्रदायतः।
ध्रुवका पतिता चित्रे वार्तिके त्वा दिमे उभे।।
ध्रुवत्ये कैव ध्रुवका कला के भेद शास्त्र और संगीत कला की क्रिया दोनों के मतानुसार क्रम में आधे-आधे हैं। अर्थात् ध्रुवका के आधे तौल के समान

सर्पिणी, सर्पिणी की आधी कृष्णा, इसी प्रकार क्रम से अन्य कलाओं की तौल कही गई है। विशेषतः कलाओं का प्रयोग नृत्य कला की क्रिया में अधिक स्पष्ट होता है। सूचना - आवाप अर्थात् एक समय में एक क्रिया को छोड़ना और दूसरी क्रिया को पकड़ना संस्कृत भाषा के साहित्य में आवाप कहते हैं।

The question now arises as to how important are these different facets of Kala for a student to learn and practice. The answer is all are equally important as long as you need those hand movements to produce the right sound. However, balance is the key to playing Indian rhythmic instruments. For example, in Mridang the alternate playing of right and left hands in the proper sequence is essential for excellent playing. If one uses the same hand on consequent beats, the outcome will not be good musically and beats will be off tune.

However, actual Musical texts have put appropriate theoretical measurement weightage on these Kala in the following terms. The weightage of each Kala halves according to ancient texts as we go down the above list. *So most important is the Dhruvka kala, meaning the accompanying excellence with Words and Vocal and the other Kalas consequently halve as we go down.* So, Sarpini is half in weightage of performance as compared to Dhruvka and Krishna is half of Sarpini and so on. All these Kalas are even more profoundly manifested in the musical form of Dance.

114

Author's Instrument of Laya Experimentation at Krishna Leela,
Tulsi Ghat, Varanasi, India

Laya (Vilambit, Madhyam, Drut)

The journey of exploring the Science and Sensibilities of Taal so far has taken us to the realm of understanding Kaal, Maarg, Kriya, Ang, Graha, Jati and Kala. Now we arrive at one of the most important facets of understanding the essence of Taal called "Laya". **Laya can be philosophically understood as that momentum which is essentially the momentum of the life force itself.** All movement in the natural world surrounding us has a particular speed which can be sensed as the Laya of the nature around us. To be in sync with the natural world is to be in Laya of life. When one becomes "out of tune" with the creation and the natural world around us, one can feel the sense of "Pralaya" which is the opposite force of Laya and is negative in nature.

From this innate understanding of Laya and its opposite in its natural form let us suffice to say that our purpose in music is Never to create anything negative. So what concerns us utmost in musical rhythm is "Laya", the constructive movement of life, within which music of universe is constantly and positively evolving.

लय विभाग

क्रिया नंतर विश्रांतीर्लय: स त्रिविधो मत:।

द्रुतो मध्यो विलंबश्च द्रुत: शीघ्र तमो मत:।।

द्विगुण द्विगुणौ ज्ञेयो तस्मान्मध्य विलंबितौ ।

मार्ग भेदाछ्रिर क्षिप्र मध्य भावैर नेकधा।।

क्रिया के आरम्भ होने के पश्चात् होने वाली विश्रांति (ठहराव) को लय कहते हैं। लय तीन प्रकार की मानी गई है द्रुत (जल्द), मध्य (न बहुत धीरे न जल्द) और विलंबित (धीरे)। दूसरा प्रकार- द्रुत लय शीघ्र होने वाली क्रिया को कहते हैं। मध्य लय द्रुत लय से दूनी विश्रांति वाली क्रिया को कहते हैं। विलंबित लय मध्य लय से दूने विश्रांति वाली क्रिया को कहते हैं। इन भेदों से एक मात्रा के भीतर भी द्रुत, मध्य और विलंबित लय काल के अन्तर्गत हो सकता है। उपरोक्त क्रम के अनुसार मार्ग भेदों से भी लय चिर, क्षिप्र और मध्य भावों से कई प्रकार की होती है। जैसे दक्षिण मार्ग में चिर भाव (स्थाई), चित्र मार्ग में क्षिप्र भाव (जल्दी) और वार्तिक मार्ग में न बहुत जल्दी न धीरे भाव (मध्य), काल के अनुसार मात्रा प्रमाण से होंगी। विश्रांति काल के एक ही समान रूप रहने पर भी उन-उन क्रियाओं द्वारा निदर्शित जैसे मार्ग भेदों से लय के भी भेद हो सकते हैं। इसी प्रकार यदि भेद किये जायें तो द्रुत में भी द्रुत मध्य, द्रुत विलंबित, मध्य मध्य, विलंबित मध्य, विलंबित मध्य, विलंबित द्रुत इत्यादि अनेक भाग होते जाएँगे, किन्तु इनका उपयोग संगीत में अनुपकारी होने से इनसे काम नहीं लिया जाता। केवल अक्षर पद और वाक्य में द्रुत, मध्य और विलंबित लय संगीत के लिए स्पष्ट उपयोगी मानी गई है।

Now let's explore the practical aspect of Laya. Once the action of creating rhythm is initiated with a physical act of striking a Mridang or Tabla, Laya is the period of interval or rest after the initial physical action that is "Kriya". The second "Kriya" then begins after the Laya interval ends. This is a very important concept and one that must be understood by a student willing to go to the depth of Indian rhythm. One might initially be mistaken to think that Laya can be created by striking and playing the drums, but that is not the case. In fact, Laya is experienced only and cannot be seen as the physical act of playing the drum. Essentially, Laya is the "Gap" of time between the consequent physical actions of playing the rhythm with hands.

An example of Moviemaking can be used here to supplement our explanation in this concept. When one makes a movie, one film the action "Kriya" and records the movie on a digital medium. However, that does not make the movie. The movie is actually made by editing and/or removing the extra unimportant portions of filmed scenes. It is the Editing between consequent pieces of actions of storytelling that makes a movie. Actual movie watcher does not see the piece of film that is edited out, but "feels" the emotion of the story.

Laya is similar in this aspect that the actual listener of rhythm can't see Laya but can "feel" the all defining aspect of rhythm in experiencing this Laya when a Musician plays their rhythmic instrument. For a sensitive musician to be in Laya is to be alive and to be out of Laya (or Pralaya) is akin to death. A listener can't see Laya but the moment the Rhythm is out of Laya and not in sync either with Vocal, or Instrument or Dance, the listener can feel the negative feeling of Pralaya. All musicians, whether beginners or masters, must strive to be continuously in Laya and avoid its opposite effect at all times. This is where Practice makes Perfect. The "riyaz" is the physical Practice and continuous study of various aspects of Taal, whereby Laya becomes a natural and spontaneous aspect of one's rhythm creation and performance.

This Laya in English transliteration has the closest but imperfect synonym of Tempo. In practical form, Laya is of three essential types.

1. Drut (fast tempo)
2. Madhyam (medium tempo)
3. Vilambit (slow tempo)

These three can be understood in experiential sense as follows. Drut is the "Kriya" or action of playing the drums at fast speed. Madhyam is action with "Double the rest of Drut time interval" between consequent "Kriya". Vilambit is the "Double rest interval of Madhyam Laya" between consequent Kriya. In other words, Vilambit is at half speed of Madhyam which itself is at half speed of Drut tempo.

Thus, even within a single "Matra" or beat, one can have multiple levels of Laya. E.g.

4 Matra Drut = 2 Matra Madhyam = 1 Matra Vilambit. This is clearly understood now from our discussion above that a single Vilambit Laya Matra will have 2 Madhyam Matra and 4 Matra of Drut Laya.

Laya and its definition do not stop here. Laya also has its sub characteristics to define many more layers of combinations of the tempo when one applies the above three Vilambit, Madhyam and Drut concepts to the "Maarg" that we discussed in the previous chapter. So based on classifications of Maarg also we have three strata of Laya;

    a. Dakshin Marg = Chir (Vilambit) Bhava (emotional sensibility)
    b. Sthayi (Chitra) Marg = Kshipra (Drut) Bhava
    c. Vartik Marg = Madhyam Bhava

Thus, when one superimposes these differences of Maarg on top of the original Laya definition of the "Interval" or gap between two consequent actions, we understand that even within Drut Laya we can have three sub-classifications of

119

Laya due to the above three classifications of Maarg. Similarly, Madhyam and Vilambit Laya based on Matra concept will have their own sub-classifications of three Maarg based Laya within it. Thus we end up with many distinct levels of Laya due to Matra and Maarg uniqueness. So, in theory, we have can have flavors of Laya such as;

1. Drut
2. Drut Madhyam
3. Drut Vilambit
4. Madhyam
5. Madhyam Drut
6. Madhyam Vilambit
7. Vilambit
8. Vilambit Madhyam
9. Vilambit Drut...etc.

However, in practical terms of spoken word and actual physical action, these Laya characteristics are extremely difficult to be demonstrated and used. So for our Musical analysis, the sages and practitioners have suggested that we must focus on three main aspects of Laya based on Matra that we discussed viz. Vilambit, Madhyam, and Drut Laya.

Yati

यति वर्णन
लय प्रवृत्ति नियमो यति रित्य भिधीयते।
समा, स्रोतोवहा चैव मृदंगा च पिपीलिका।।
गोपुच्छा, चेति विबुधैर्यति: पञ्च विधो दिता।
यति लय की प्रवृत्ति अर्थात् चाल क्रम (गति) के नियम को कहते हैं। यति, पाँच
प्रकार की मानी गई है। समा, स्रोतोवहा, मृदंगा, पिपीलिका और गोपुच्छा।

Having understood what "Laya" is, we need to now understand how the Laya actually (even though it can't be seen) manifests itself in music. **This is through the most unique concept of "Yati" which is defined as rules of the practical form (or rules of the actual style of walking and unfolding) of the evolution of a particular Laya when a Taal is being played.** When a rhythmic structure or phrase is played, it has three sections, Beginning (Adi), Middle (Madhya) and End (Ant). The Laya combinations in these sections define their Yati styles.

1. Sama Yati
2. Strotovaha Yati
3. Mridanga Yati
4. Pipilika Yati &
5. Gopucha Yati

Let us now analyze each Yati and its formation and nature. Analysis of each Yati confirms to us the deep natural laws of universal music. The sages of Rhythm science in India have taken inspiration from nature in all aspects in defining these formations of rhythmic tonalities that when combined with Yatis produce a unique pattern of Laya sequences depicting the beauty of nature. **Music, after all, is a universal language based on irrefutable Natural laws in existence.**

NOTE: *With each explanation below we now enter into more advanced descriptions of the concepts where knowledge of Hindi language is required to further understand the practice of implementing these Taal Yatis with examples given in the Hindi text that can't be transliterated into English in this volume. Hence for the seeker, we assume some basic appropriate knowledge of Hindi and thirst for practicing and applying these ideas on their own musical rhythmic instruments to "feel" and experience these ideas in practice.*

Sama Yati

आदि मध्या वसानेषु लयै कत्वे समा त्रिधा।

आदि मध्य और अन्त इन भेदों से समा, यति भी तीन प्रकार की होती है। आरम्भ, बीच और अन्त तीनों स्थानों पर बराबर एक ही की लय होना यही समा, यति का रूप है। समा, यति का सूक्ष्म रूप आरम्भ ।, बीच । और अन्त ।, तीनों स्थान पर गति समान हो।

समा, यति का स्थूल रूप-

```
  +              o              ।
  १ (आदि) २    ३    ४    ५ (मध्य) ६
  धागेतिट   धागेतिट तागतिट तागेतिट धागेतिट  किटधागे

              ।              ।
  ७      ८      ९      १० (अन्त) ११  १२ १  ० ३
  तिटाकट तागेतिट किटतागे,  तिटकिट   गदिगनधा (धा, धा, धा,)
```

Sama in Sanskrit refers to the meaning of the same or equal in English. Also, Sama refers to a lake or a river with steady water flowing in a calm manner and hence the Laya aspects in the **Sama Yati** recreate the steady flow of calm rhythm. Here in this Yati format, the three parts of the rhythmic structure Beginning, Middle and End, the Laya is the same or equal. A Taal formation or composition where the three parts Beginning (Adi), Middle (Madhya) and End (Ant) of that rhythmic structure sections all have exactly same Laya (can be any Laya, Drut, Madhyam or Vilambit as long as it is same for all sections) is called Sama Yati.

Sama Yati = (Adi, Madhya and Ant parts of Rhythmic structure) = Same Laya

Strotovaha Yati

चिर मध्य द्रुत लयैर्युक्ता स्रोतोवहा मता।
अन्या बिल्व मध्याध्यां मध्य द्रुत वती परा।।
स्रोतोवहा जो क्रम से आदि में विलम्बित लय, बीच में मध्य लय और अंत में द्रुत

लय हो उसे स्रोतोवहा कहते हैं।

           (आदि) (मध्य) (अन्त)   १(आदि)२
स्रोतोवहा का सूक्ष्म रूप S, SS, ।।।, स्थूल रूप- धाकिट तकिटत,
३ ४ (मध्य) ५ ६ ७ ८ (अन्त)
काकिट, धाकिटतक धुमकिट तक, धुमकिट तक था किटधुम किटतक,तकिटतका
                      ९   १० १२ ३
किट तक् गदिगन था. किटतक् गदिगन्धा किटतक् गदिगन् था । धा, किट, धा धा था
था,
सूचना - कुछ संज्ञीतज्ञ विद्वानों का मत है कि स्रोतोवहा में बीच की लय
विलम्बित, आदि में मध्य, लय और अन्त में द्रुत लय हो स्रोतोवहा कहते हैं।

   (मध्य)(विलम्बित) (द्रुत)   १ (आदि)   २(मध्य)
सूक्ष्म रूप- S, S, ।, स्रोतोवहा का स्थूल रूप- धातिरकिटतक, कतथेन,
३ (अन्त) ४
किटतकधा किटतकधा किटतकधा ।

Strotovaha Yati refers to a bubbling spring and a body of
water with gradual crescendo tone and flow of spring or a
river. This Yati also depicts a falling waterfall and its
crescendo on top and then calmness after falling. Another
inspirational analogy with this Yati is that it also represents a
flow akin to the flow of the Vedic Strotras with the verbal
sound of the Strotra (compositions). Thus, it has a gradual
crescendo formation as follows;

Strotovaha Yati -   *Adi section = Vilambit Laya*
                            Madhya section = Madhya Laya
                            Ant section = Drut Laya

Some Scholars have also defined Strotovaha Yati as follows;
(both can be used in practice but more common form is one
above)

Strotovaha Yati -   *Adi section = Madhya Laya*
(Alt. version)   *Madhya section = Vilambit Laya*

Ant section = Drut Laya

Mridanga Yati

Mridanga Yati is named after the oldest Rhythmic Indian instrument "Mridang" and is based on the shape of this drum which is wide on one end, wider in middle and then tapering on the other end. Thus it represents a flow of Laya in this Yati akin to the shape of Mridang itself.

मृदंगा, तु द्रुता द्यन्ता, मध्ये मध्य लयान्विता ।
तथै वान्या द्रुता द्यन्ता, ज्ञेया मध्ये विलम्बिता ।।

मृदंगा आदि और अन्त में द्रुत लय, बीच में मध्य लय कुछ अन्य विद्वानों के मतानुसार आदि अन्त में द्रुत लय और बीच में विलम्बित लय ।

मृदंगा आदि और अन्त में द्रुत लय, बीच में मध्य लय कुछ अन्य विद्वानों के मतानुसार आदि अन्त में द्रुत लय और बीच में विलम्बित लय ।

मृदंगा पहली स्थूल रूप-

१   (आदि)   २ ३(मध्य) ४   ५(अन्त)

कातिरकिट धिरकिट तक् तिरकिट धा, दिनूदिन् तिटतित कातिरकिट तिरकट

          ६

धिरकिट तक् धा ।

मृदंगा दूसरी -

१(आदि)   २   ३ ४(मध्य)

तिरकिटतक् धिरकिटतक् तिरकिट धिरधिरकिटतक् धातिरकिटतक् , धा, धा,

५ ६ ७ (अन्त)   ८   ९

तू , ना कत्तिरतक् धा, कत्तिर किटतकधा, कत्तिरकिट तक् धा ।

मृदंगा तीसरी, मध्य और आदि में द्रुत, लय अन्त में विलम्बित लय इस प्रकार मृदंगा के तीन भेद हैं।

Mridanga Yati -   *Adi section = Drut Laya*
                          Madhya section = Madhya Laya
                          Ant section = Drut Laya

Some Scholars have also defined Mridanga Yati as follows;
(both can be used in practice and are equally prevalent)

Mridanga Yati -   *Adi section = Drut Laya*
(Alt. version)   *Madhya section = Vilambit Laya*

126

Ant section = Drut Laya

१ (आदि)   २ (मध्य)
स्थूल रूप- किटतक गदिगनधाकिटतक , गदिगनधा किटतक गदिगन,
            ३ (अन्त) ४   ५   ६
            धागेतिट, तागेतिट, किड़धातिट, गदिगन ।

Pipilika Yati

पिपीलिका गती अर्थात चींटी की चाल

पिपीलिकातुकथिता, मध्येद्रुत विलम्बिता।

अद्यन्तमध्याचैवान्या, प्रोक्तामध्येद्रुतान्विता ।।

मध्ये मध्यान्विता द्वन्त विलम्बित लयापरा ।

जो लय आदि अन्त में विलम्बित और बीच में द्रुत लय रहती है। उसे पिपीलिका कहते हैं। ऐसी ही कल्पना करके कई विद्वान आदि अन्त में मध्य लय और बीच में द्रुत लय तथा कोई लोग आदि अन्त में मध्य लय और बीच में विलम्बित लय मानते हैं।

१(आदि) २ ३ (मध्य) ४ ५

पिपीलिक पहली- धातिरकिटतक, ताऽऽ, धिन्तरान्धात्नाकत्, कतिधा, ऽनतिट,

१(आदि) २ ३ ४ ५(मध्य) ६

पिपीलिका दूसरी- धागे तिट, तागे तिट, धगेनघकिटधागे, तिटकिट किटतक, (अन्त)

७ ८ ९ १० ११ १२ १३ १४ १५

किट तक धाऽ किट तक धाऽ किट तक धाऽ ।

पिपीलिका तीसरी- (आदि) (मध्य) (अन्त)

१ २ ३ ४ ५ ६७ ८ ९ १०

किट तक ताऽ, किटतक ताऽ, धा ऽ दिन् ता, क ह्राति

११ १२ १३ १४ १५ १६ १

ह्राऽ क ह्राति ह्राऽ क ह्राति ह्रा ।

Pipilika Yati is named after the Sanskrit word for Ant Hill and the movement of ants in a specific flow. **Ants have a much-disciplined way of formations in movements with groups performing tasks in the beginning as well as end collectively and in middle moving alone a well-formed slim and slender path of individual ant movements fast or slow based on work needing to be performed.** An ant is one of the smallest beings but its commitment and dedication to its work is the paramount example of professionalism. Thus, inspired on the ant's structured movements from nature, this Yati bulges on both ends (collective ant movements) with the middle being slender (individual pause and movement action of ants in a single file) in the first two flavors below and opposite variation in third.

Scholars have defined three distinct flavors of this Yati as

follows; (all can be used in practice and are equally prevalent)

Pipilika Yati (a) -   *Adi section = Vilambit Laya*
                              Madhya section = Drut Laya
                              Ant section = Vilambit Laya

Pipilika Yati (b) -   *Adi section = Madhyam Laya*
(Alt. version)   *Madhya section = Drut Laya*
                              Ant section = Madhyam Laya

Pipilika Yati (c) -   *Adi section = Madhyam Laya*
(Alt. version)   *Madhya section = Vilambit Laya*
                              Ant section = Madhyam Laya

Gopucha Yati

सूचना - समा यति अर्थात् जहाँ पर जल का प्रवाह स्थिर गति के साथ बहता हो समा यति कहते हैं। स्रोतोवहा जैसे झरने का प्रवाह, मृदंगा मृदंग के आकार के सामान पिपीलिका चींटी के गति के सदृश और गोपुच्छा अर्थात् गऊ के पुच्छ के समान हैं और भी अनेक प्रकार की यति यथा लय की गति होती है, उनका विवेचन आगे किया गया है

यति वर्णन

मध्या रम्भा विलम्बान्ता, गोपुच्छातु यतिर्मता ।

गोपुच्छा- जो गति मध्य लय से आरम्भ होकर क्रमशः विलंबित लय होती जावे ऐसी गति को गोपुच्छा यति कहते हैं।

१(आदि) २   ३   ४   ५

गोपुच्छा पहली, स्थूल रूप - धाकिट तकिटत काकिट धुमकिट तकिटत

६   ७ (मध्य) ८ ९(अन्त) १० ११ १२

काकिट दिन   S, ता S S धा

गोपुच्छा दूसरी - द्रुत मध्य विलम्बैस्याग्दोपुच्छा, द्रुत मध्यभाक ।

गोपुच्छा दूसरी और तीसरी जो लय की गति क्रमशः द्रुत, मध्य और विलम्बित होती जावे उसे भी गोपुच्छा कहते हैं अथवा आरम्भ में द्रुत और मध्य लय की गति क्रमशः होती जावे उसे भी गोपुच्छा कहते हैं।

१ (आदि)   २(मध्य) ३   ४(अन्त)५

गोपुच्छा दूसरी, स्थूल रूप - धिनतिरकिटतकता, कतधेन नागेतिट, धादि म्

६   ७

ता   धा

१   २   ३

गोपुच्छा तीसरी, स्थूल रूप - कत्तिर किट तकधा, घेघेतिट धागेतिट,

४   ५

गदिगन नागेतिट

Gopucha (Cow's Tail) Yati is named after the Sanskrit word for Cow's Tail and its shape which curves at the end. Cow is a sacred being akin to Mother in Indian ethos and hence this Gopucha Yati is also using Cow's tail as an inspiration for its flow.

Scholars have defined three distinct flavors of this Gopucha Yati as follows; (all can be used in practice and are equally prevalent)

Gopucha Yati (a) -   *Adi section = Madhya Laya*
                                    Madhya section = Madhya Laya
                                    Ant section = Vilambit Laya

Gopucha Yati (b) -   *Adi section = Drut Laya*
(Alt. version)   *Madhya section = Madhya Laya*
                                    Ant section = Vilambit Laya

Gopucha Yati (c) -   *Adi section = Drut Laya*
(Alt. version)   *Madhya section = Madhyam Laya*
                                    Ant section = Madhyam Laya

Lord Krishna's favorite Cows (inspiration for Gopucha Yati)....resting under their favorite Goverdhan Hill in Brij Bhoomi, India

Dugun, Savai, Dedhi Yati

While the main five Yatis have been described above, the below are some more Yatis that are purely based on variations of mathematical tempo calculations.

In **"Dugun Yati"** as the name suggests, the Laya flows at double speed. So the interval between several pieces of rhythmic Taal structure remains the same but the number of Matras representing the Rhythmic sounds is doubled in the same space.

दूनी यति - मात्रा द्वै गुण्य गत्या श्चेत् ।
द्रुत   मित्य   भिधीयते ।।
सूचना - मात्रा जब दूनी गति के साथ होती है अर्थात् स्थान स्थाई ही रहता है। किन्तु मात्रा की गणना दूनी होती है ऐसी गति को द्रुत (दूनी लय कहते हैं)

ताल चौताला के दूनी लय का बोल

+		o		I		o	
१	२	३	४	५	६	७	८

धागेतिट धागेतिट तागेतिट तागेतिट किड्धाकिट धागेतिट गादिगन नागेतिट

I			I		+	
९	१०	११	१२	१		

कतिटत गेनधा किटतगे नधान धा
१ २ २ ३

सवाई, यति- चतुर्थांश श्चैक मात्रा मिलिता यति मास्थिता ।
        लयस्य विकृति जाता सपादैक उदाहृता ।।
मात्रा और मात्रा का चौथाई हिस्सा दोनों मिलकर जो गति होती है उसे सवाई लय कहते हैं।
ताल चौताला के सवाई लय की गति का (स्थूल रूप) -

+			o			I			o			I			I
१	२	३	४	५	६	७	८	९	१०	११	१२	१३	१४	१५	

धा तक तिट किट तक तिट किट तक धा दिन् तग तिट किट तक   धा

The same approach above can also be recreated when at the same time interval, instead of double Matras, we play 1.25 Matras. It's called **"Savai Yati"** representing the Savai Laya or 1.25 Laya speed of rhythm. Savai in Hindi means 1 and ¼ (1.25)

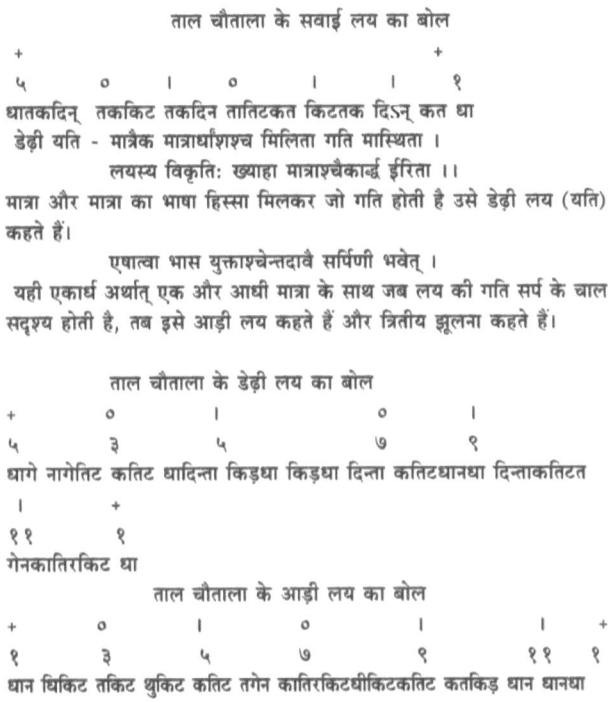

ताल चौताला के सवाई लय का बोल

+ +
५ ० । ० । । १
धातकदिन् तककिट तकदिन तातिटकत किटतक दिऽन् कत धा

डेढ़ी यति - मात्रैक मात्रार्धांशश्च मिलिता गति मास्थिता ।
लयस्य विकृतिः ख्याहा मात्राश्चैकार्द्ध ईरिता ।।

मात्रा और मात्रा का भाषा हिस्सा मिलकर जो गति होती है उसे डेढ़ी लय (यति) कहते हैं।

एषात्वा भास युक्ताश्चेन्तदावै सर्पिणी भवेत् ।

यही एकार्ध अर्थात् एक और आधी मात्रा के साथ जब लय की गति सर्प के चाल सदृश्य होती है, तब इसे आड़ी लय कहते हैं और त्रितीय झूलना कहते हैं।

ताल चौताला के डेढ़ी लय का बोल

+ ० । ० ।
५ ३ ५ ७ ९
धागे नागेतिट कतिट धादिन्ता किड़धा किड़धा दिन्ता कतिटधानधा दिन्ताकतिटत
। +
११ १
गेनकातिरकिट धा

ताल चौताला के आड़ी लय का बोल

+ ० । ० । । +
१ ३ ५ ७ ९ ११ १
धान धिकिट तकिट थुकिट कतिट तगेन कातिरकिटधीकिटकतिट कतकिड़ धान धानधा

When in the same space as discussed above in this Chautaal example we play 1 and ½ Matras (1.5), it becomes **"Dedhi" Laya Yati**.

And finally, if we play in the same time-space of 1.75 Matras, it becomes **"Paun Dugun" Laya**. Paun Dugun means 1.75 in Hindi.

पौने दूनी, यति = मात्रा श्चैकोन पदांश सहिता गति - मास्थिता ।
यतिर्लय, समा योगात् एको नांशान्मि - ईरिता ।।
मात्रा और मात्रा का पौन हिस्सा मिलकर जो गति होती है उसको ऊनांश कहते हैं
अर्थात् पौनें दूनी- लय कहते हैं।

ताल चौताला के पौने दूनी लय का बोल

+		o				o			
९		३	५	६	७		९		

धाकिट ताकिट काकिट किटतकधुम किटतकि टतकाकिटकिट ताकिटत

		+	
११		९	

कातिटगादिगन धा ।

Prastar

प्रस्तार–नियम

न्यस्याल्प माद्यान्महतो धस्ताच्छेषं यथोपरि ।
प्रागूने वाम संस्थास्तु संभवे महतो लिखेत् ।।
अल्पान संभवे तालपूर्त्यैं भूयोप्ययं विधि: ।
सर्वाण्व वधिरा लेख्य: प्रस्तारो यं द्रुते लघौ ।।
गुरौप्लुते समस्ते च व्यस्ते व्यस्ते त्वणो नस: ।
एवं तालस्य विबुधैर्दश प्राणा निरुपिता: ।।

जिस किसी ताल का प्रस्तार (रूप) लिखना हो उसका पहले छोटे से छोटा रूप लिखें अर्थात् पहले कहे हुए तालों के अंग में छोटे से छोटा स्थूल रूप अणु (̆) का ही माना गया है, जो एक मात्रा का चतुर्थ हिस्सा होता है। उससे प्रारम्भ करें। पश्चात् क्रम से बड़ा-बड़ा रूप, अर्थात् लघु (।) गुरु (S) प्लुतादि (ऽ)का रूप निर्माण करते हुए कोई भी ताल के प्रस्तार का रूप निश्चित कर सकता है ।

यदि इस क्रम के अनुसार ताल पूर्ति होने में कुछ बाकी रह जाए तो ताल के अंगों में से बड़े रूप को अर्थात् लघु गुरु प्लुतादि से लिखना आरम्भ करें। फिर भी ताल का प्रस्तार बनाना असम्भव मालूम पड़ता हो तो अणु (̆) द्रुतादि (०) को रखते हुए ताल का प्रस्तार निश्चित करें।

In this final Prana of Taal, we now identify the techniques of "Prastar' (Expanding) the nature of the Taal in playing formations. This Prastar is the final most important element of Taal Prana because after understanding all the nuances of this science in this final act of playing one is "presenting" their art to the listener and their soul for their own self-satisfaction. In Indian Classical Music, there are many forms of Prastar namely, Swar Prastar (expanding melody using the seven notes in a given Raga), Laya Prastar (expanding the Taal and its Laya in many unique mannerisms), Theka Prastar (expanding upon base Taal words), Kaayda Prastar (expanding on further detailed aspects of a Taal in Tabla), Paran Prastar (specifically in Mridang, expanding on specific poetic formations of Shashabd Kriya of Taal and its accompanied rhythmic intonations).

In simple words, Prastar is basically an act of modifying the same basic Bols of Taal and repeating them with varied creativity at one's own pleasure for the length of time that the artist wishes. This is the foundational definition of Prastar. In a solo performance, this becomes of paramount importance.

Now let's understand some technical methods of Prastar. There are two methods of creating Taal Prastar in the Indian system. The first technique is as follows;

1. Whichever Taal is being played as a solo Taal, first identify the smallest component of that Taal. The smallest component of Taal is identified as "Anu" ˘ symbol and signifies ¼ of Matra.
2. Start with Anu Matra and sequentially increase the size of Matras in the composition to Laghu (1 Matra) – symbol. |
3. Guru (2 Matras) – the symbol ∫
4. Plut (3 Matras)
5. If even after this the Taal still is not completely expanded upon, then start with bigger Taal Ang formations described earlier and repeat until satisfied with expanding on a rhythm for a particular   Taal.

प्रस्तार का दूसरा प्रकार

जिस ताल का प्रस्तार लिखना हो उसकी मात्राओं को गिने और अणु मान कर लिखे। ताल के अनुसार क्रमशः लघु, गुरु और प्लुत रखता जाए। इस प्रकार लघु, गुरु और प्लुत के प्रस्तार की विधि है। चाहे ऊपर लिखे हुए ताल का एक ही अंग लेकर हो अथवा इन एक के साथ ताल का दूसरा अंग भी मिलकर प्रस्तार होता है । किन्तु एक अणु या एक द्रुत में होना असम्भव है। यदि एक लघु का प्रस्तार लिखना हो तो पहले एक लघु लिखें पश्चात् अणु, द्रुत इत्यादि लिखता जाए। ऐसा करने पर यदि प्रस्तार असम्भव जान पड़े तो उक्त ताल के अवयवानुसार भाव लेकर लिखते जाएँ। इसी प्रकार गुरु और प्लुत इत्यादि की क्रिया होती हैं।

उदाहरणार्थ मान लो चौताल का प्रस्तार लिखना है और ताल के अंग में सबसे छोटा रूप अणु होता है। अतएव चौताल में १२ अणु हुए, चौताल में पहला और दूसरा ताल ४ अणु का तीसरा और चौथा ताल २ अणु का है, अतः इसका

प्रस्तार इस प्रकार लिखा जाएगा --

पहला, ४ अणु = १ लघु ।
दूसरा, ४ अणु = १ लघु ।   यह चिन्ह उपरोक्त परिभाषा के अनुसार
तीसरा, २ अणु = १ द्रुत ०   एक के पश्चात् एक लिखे जाएँगे ।
चौथा, २ अणु = १ द्रुत ०

इस प्रकार ताल के अंग में से बड़ा रूप लेकर छोटे रूप को संग रखते हुए चौताल का प्रस्तार रूप बना है।

चौताल का प्रस्तार रूप - लघु २, द्रुत २, ।।००

दूसरा प्रकार - ताल के अंग में से केवल एक ही बड़ा रूप लेकर, सदानन्दः, एक ताल का रूप बना है। जिसमें तीन मात्रा मानी जाती है। एक ताल, वा सदानन्दः ताल में, एक लघु है तो त्रस्य जाति का माना जाता है और इसमें ताल एक ही है। सदानन्दः, एक ताल का प्रस्तार रूप लघु।

The second way of Prastar of a Taal is whereby

1. Each Matra is considered the smallest "Ang" of a Taal i.e. Anu. So Chautaal 12 Matras has 12 Anu angs.
2. Then group these Anus in sections or pieces of the Taal and based on groupings, identify Laghu Guru Plut, etc.
3. In the case of 12 beat Chautaal, we will see that the 4 sections have 4,4,2,2 sections composed of 12 beats. The first section and second section have 4 anus each equaling 1 Laghu.

137

4. The third and fourth pieces of Chautaal have 2 anus and that equals 1 Drut from our Ang discussion earlier.

5. Thus we conclude that the final form of Chautaal is Laghu, Laghu, Drut, and Drut representing 4,4,2,2 beat formations.

This concludes our analysis of the Ten Prana of Taal. To summarize let's think for a moment of the fact that THE "Providence" created our human bodies and then "infused" life in our bodies through the "Prana".

In precisely the same manner Music and its entire Musical Body is "infused" with the life force only though "Taal" which is now clear in all readers of this reference as the "Prana" of all universal music. Without Taal, music would be a lifeless body. *It is ONLY with Taal that Music comes alive.*

Finally, if Taal is the Prana of Music the true seeker also strives to understand the inner Ten Prana of Taal itself. *These Ten Prana of Taal and residing within itself are what define Taal as its foundational life force of Music.* It is the Author's humble hope that even if some small appreciation of the Ten Prana of Taal the reader would obtain from this reference volume, the Author's efforts would have been worthwhile. "Saarthak". Understanding these ideas coupled with the physical act of committed practice would surely take one on the path of liberation in this journey of Laya.

Section 3 - Practical Concepts

Tihai, Mohra, Bol, Paran, Tukda

In this section, we now switch from analysis of the science, quantification and qualitative origins of the Taal Shastra to the practical know-how. Today we hear many concerts and many listeners go to these but they are just skimming on the surface of what is going on. Practically speaking if the listener and practitioner understand meanings of the concepts that the learned experts and "Vidwan Gunijan" (Learned Taal scholars) use in their performance our experience and enjoyment of this process would multiply infinitely. This section hence focusses on learning some of the terms used by experts in their day to day practice of applying earlier science of Taal in the actual act of artistic creation and playing.

These few chapters bring a curious student up to speed with the precise meanings of these ideas and terms. Once we understand their meaning, one can apply a better mindset in the practice of active listening and playing some of these structures. The following practical chapters and the items written about in the advanced Chand section thus assume that the reader knows "How to play" the rhythmic instrument but the focus here is on understanding "What is the mechanism and logic behind what is being played".

Tihai

"Tihai' – in Hindi means three times.

तिहाई, मोहरा, गत, बॉट, बोल, परन, टुकड़ा इत्यादि की परिभाषा का विवेचन तिहाई
कुछ अक्षरों के नियत शब्दों को तीन बार क्रिया करके सम पर आने को तिहाई करते हैं।
यथा:- तिटकत गदि गन धा तिट कत गदि गन धा तिट कत गदि गन धा ।

Tihai is the main fundamental idea in Taal Shastra. Its origins go back to the very meaning of 'Learning' and studying something. **The Power of Three** (Trik in Sanskrit) is a universal creative concept that we have recognized scientifically and it is applied by neuroscientists in learning techniques. This concept of geometry and mathematics is described at length in the second volume of this Nada Yoga Series 'Science of Melody'. For the reader who is really curious to understand the reasons behind why Tihai are important in music, the chapter on sense and sensibilities of Melodic system describes the geometrical principles of music and discusses this in detail in the companion reference book for the present volume. *That which is important to learn must be repeated in practice. (This is the essence of the Sanskrit word "Abhyaas")* Since ancient Vedic times, the chants and learning systems use three repetitions as a must to solidify something in our minds. The more iterations of three, the better one's practice.

Tihai idea derives from this root understanding. A bol or phrase is played three times prior to arriving at the Sama (which is a Graha Prana of Taal). This Sama is the Union of Taal and non-rhythmic portions of Music and hence the arrival here with a Tihai puts a mesmerizing effect and weight of the moment of Sama on the listener. There is even a separate scientific method of how to create a Tihai in any Taal variation. This advanced technique of creating a Tihai is beyond our current topic but would be covered in future writings on the practical aspects of "How to Play Mridang".

Mohra

मोहरा

लय की गति के साथ जिन शब्दों की भी क्रिया आरम्भ करके पुन: सम के साथ मिलाते हैं ऐसी क्रिया को मोहरा करते हैं, किन्तु गाने में स्थाई जिसे कहते हैं चाहे ख्याल अथवा धुरपद की हो जिन शब्दों से आरम्भ करते हैं उन्हीं शब्दों को लेकर पुन: सम पर आते हैं ऐसी क्रिया को मोहरा कहते हैं परन्तु ताल अध्याय में बोल आरम्भ करके पुन: कोई भी शब्द लगाकर सम पर आने को मोहरा कहते हैं।

१	२	३	४	५	६	७	८

यथा :- धागे न धा तिट घिट तरान क S घ्री S S थुं S न् त था

Next, we discuss the concept of "Mohra", which in English means Face or sometimes even Mouth. **The reason this name is utilized is that just as our face is the starting point of one's visual frame of reference in starting any conversation and then you go to other subjects, in Taal auditory sense, the Mohra is the action of arriving at the SAMA by playing some phrases of Taal in a particular Laya prior to the starting point of non-rhythmic components of music.** This is a similar concept to "Sthayi" in the non-rhythmic aspects of Indian classical music. In Vocal music be it Dhrupad format, Khayal or even devotional Kirtan, one sings a "Sthayi" (reference) phrase and then improvises the tonalities of the Raga while revisiting the Sthayi again and again and restarting the new variation from Sthayi.

In Taal practice, Mohra is analogous to Sthayi in vocal singing music. The only difference is that the Mohra gives flexibility to the Taal vadak (Player) to use any phrases while maintaining the Laya and then coming to Sama. In Vocal, the Sthayi is restrictive in that the same words and phrases must be treated as Sthayi. In Taal, Mohra can differ based upon the mood of the Taal player and the accompanying theme of music.

Bol

बोल

जिन शब्दों की गति स्याही के शब्दों से अधिक सम्बन्ध रखते हुए कुछ दूरी को लेते हुए एक सी अथवा भिन्न प्रकार की चाल व शब्दों का परिवर्तन करते हुए लय के साथ, तिहाई के साथ या बिना तिहाई के सम पर आवे ऐसे शब्दों की रचना को बोल कहते हैं।

१ २ ३ ४ ५ ६ ७
यथा :- धागेतिट धागेतिट तागेतिट तागेतिट कृधाकिट कृधाकिट घिंतिरकिट तक
८ ९ १० ११ १२ १
तागेतिट धतगे ऽ त्र कृतत् धेत्ता गदिगन धा ।

Bol – is the main signifier meaning (words) in English. These are the words of the Taal language. **These are the phrases that are created, composed and played using the science we discussed above and applied to other aspects of non-rhythmic music to create an overall harmony of rhythm, vocal and instruments as well as dance.** The Taal phrases in Bol are more related to the right hand "Siyahi" (black ink) section of Mridang or Tabla keeping a specific idea in mind in their usage or gait of playing. Bol can be ending with a Tihai or without Tihai, but like all musical ideas, MUST end at Sama.

Paran

परन

तबले के आदि वर्णों अथवा और अक्षरों को लेते हुए क्रिया करने को परन कहते हैं। परन प्रायः बड़ी बन्दिश होती है।

१ २ ३ ४ ५ ६ ७ ८ ९ १० ११ १२
यथा :- धा तिट धाधा धिन गिन तूना कत्ता धार्दिंऽना गिन धिट तिट
१ २ ३ ४ ५ ६ ७ ८ ९ १० १२ १
धिरकिट गिगिऽना किट धातीं धा तूना तक तक तक धाताधा

Paran is basically the most ancient form of Bols and these are poetries in rhythm. Parans originally came from the world of Mridang and then were adopted by Tabla. Basically a Paran is an extended set of Bols. If a Bol is comprising of one sentence of a Taal idea, Parans can have several 2, 3, 4, even more than 100 lines in them (in rare cases). Most Parans can have 30 to 40 lines on the long side and 3 to 4 on the short side. These are complex numerical and poetic combinations of patterns of Taal which often can be harmonized with actual poetry and their "Kavit Pada" in Hindi or in "Bhasha – Brij Bhasha". All of this, of course, is memorized as in Indian Music there is no practice of seeing the notations and playing. The bar is set high to use the brain cells even more vigorously by memorizing the creations and compositions.

Parans can be of many sub-varieties depending on their structure. Because they are bigger compositions, there is more possibility to engineer different flavors. Viz.

1. Chakradhar Paran
2. Farmaishi Paran
3. Sadharan Paran
4. Namaskari Paran
5. Kamali Chakradhar Paran, etc.

Tukda

टुकड़ा

जिन शब्दों की गति की चाल खण्ड करती हुई सम से उठकर तिहाई देती हुई अथवा बिना तिहाई के सम पर आए ऐसी क्रिया को टुकड़ा कहते हैं (चाहे वह गत बोल इत्यादि का हो) वही चीज़ आरम्भ से तीन या नौ बार कहन का चक्करदार टुकड़ा कहते हैं।

१ २   ३ ४ ५   ६ ६ ८ ९ १०   ११ १२
यथा :- धातिर किटतक ता धा तिर किटतक ता धि ता धि त गि ऽना धा
१३ १४   १५ १६ १
तिरकिट तक ता ऽधा ऽ न धा ।

"Tukda" – means Piece in English. In a Tukda, one uses the "Khand Jati" phrase breaking the Bols in pieces and then coming to Sama either with or without a Tihai. The Laya and flow of Bols are not smooth but rather broken into pieces. Tukda can be effectively used to migrate from one Laya to another and so on. If the same Tukda is played from Sama three times or nine times and then comes back to some it creates what is known as a "Chakradhar Paran"

Pallu, Chaupalli, Angushtana, Farad

Pallu

पल्लू

जिन शब्दों की गति की चाल बिना खण्ड किए तीन बार कहकर सम पर आए,
ऐसी क्रिया को पल्लू कहते हैं।

१   २   ३   ४   ५   ६ ७   ८   ९   १०   ११   १२
यथा :- धाधा धिता धित्ता तिटकत गदिगन धा तिटकत गदिगन धा तिटकत गदिगन धा

Pallu is a concept where one without breaking the Laya Gati
and its gait, uses a Phrase and says it (plays it) Three times and
comes to Sama. It is a type of Tihai naturally. So it flows. The
difference between a normal Tihai and Pallu is that in Tihai
one can use "Khand Jati" small pieces of bols to break the
Laya and still come to Sama.   Whereas in Pallu the gait of
Laya and its Gati is NOT broken.

Chaupalli

चौपल्ली

चौपल्ली दो प्रकार की होती है (१) बोल एक ही हों खण्ड उसके चार-चार मालूम
होते चले जाएँ । (२) बोल एक ही हों मात्रा की गति ठा, दुगुन, तिगुन, चौगुन
अथवा ठा दुगुन चौगुन हो ।

१३५.   १   २   ३   ४   ५   ६   ७   ८
उदा०-(१) धाकिट धाधिंता किड़धा धिन्तातक। कातिट धाधिन्ता किड़धा
धिन्ताकत।

९   १०   ११   १२   १३   १४   १५   १६
किड़धा-धाधा-धाधा कतिट ताऽन । किड़ ता ता ता ता ता धाकिट धाऽन ।

१   २   ३   ४ ५ ६   ७   ८ ९   १०   ११
(२) तिट कत गदि गन धा तिटकत गदिगन धा   तिटकतगदिगनधा   तिटकत
      १२
गदिगन धा ।

"Chaupalli" means a formation of four. "Chau" in Hindi means
four in English. This is a form of Bol composition where there
are two flavors as seen here;

Chaupalli 1 – Here the Bol is the same but the "Dha" and "Ta" variations are used to create groupings of 4 Matras. Hence the name Chaupalli.

Chaupalli 2 – Here the Bol is exactly same but in the 4 pieces the Laya is changed with Matra numbers changing according to Thah (1.5 tempo), Dugun (Double tempo), Tigun (Triple Tempo) and Chaugun (Quadruple tempo).

Or it can be three sections of Thah, Dugun and Chaugun.

Angushtana

अंगुश्ताना

यह शब्द फारसी का है। जो शब्द उंगली से निकलने वाले टाँकी का बर्ताव करते हुए स्याही पर भी केवल उंगली से ही निकलते हैं ऐसी क्रिया को अंगुश्ताना कहते हैं।

यथा :- तक धिन गिन धिन नादिन्ना धिन गिन धातिन् ऽ ताधिन गिन धिन् तधि नकधिन धागे नागे धिने गिन ।

"Angushtana" is a Farsi word that also illustrates that Music as a Universal language always finds a way to create new synthesis from differing cultures. During the period in Indian history when invaders from Persia and Mughal geographies came to India, the musical forms evolved in a new direction and around this time Tabla was also invented by splitting Mridang into two parts. The word Angushtana refers to our fingers and this technique is the Bols produced just by using the tips of fingers and not the whole palm in Mridang as is usually the case. In Tabla, fingers are used heavily but some Parans in Mridang can also be produced in a beautiful manner with Angushtana type of Bols. The fingers usually strike the middle part of "Siyahi" or the "Chaanti" on the border.

146

Farad or Ekkad

फरद अथवा एक्कड़

प्रचीन विद्वान प्राय: बन्दिशों की रचना जोड़ से करते थे । जिस बोल अथवा गत का जोड़ नहीं बनता था उसको फरद अथवा एक्कड़ कहते हैं।

Farad or Ekkad refers to Parans, which instead of flowing with the overall flow of the composition (Gat), stand on their own and are played not as a part of a Gat composition together with it. Hence the name Ekkad (Ek means single in English).

147

Gat, Baant, Pench, Rela, Ladi, Laggi, Uthaan

Gat

"Gat" – derives from the Sanskrit Gati meaning flowing towards a direction. **Gat in context of Taal science is a composition in Mridang and Tabla where more than usual phrases or Bols are used and instead of a single Laya, many different Laya are composed in the same composition.** The alternate "Dha" and "Ta' are used to expand upon the beginnings of Bols. Tihai is usually NOT used in a Gat.

Baant

बाँट

तबले की थोड़े से शब्दों की बन्दिश जो स्याही और टाँकी पर लय परिवर्तित करते हुए बजाई जाती है। इसमें खाली भरी का प्रयोग होता है अर्थात् एक बार धा के साथ और एक बार ता के साथ बजाया जाता है।

१   २   ३   ४   ५   ६   ७   ८   ९   १०
यथा :- धीना ऽध्रा तिर किट धीना तिरकिट धीना ऽध्रा तिरकिट तीना ऽता
११   १२ १३   १४   १५ १६
तिरकिट धिन तिरकिट धीना ऽध्रा तिरकिट ।

This is another form of Bol composition used in Tabla mainly whereby length is shorter than Gat and the same "Dha" "Ta" alternate structure is used. Here the emphasis is on Laya difference between the words played on "Siyahi" as opposed to the words on "Chaanti".

Pench

पेंच

बाँट पेश्कार अथवा कायदे की बन्दिशों को अक्षर बदल बदल कर विभिन्न लयों में उनके
बार बार बर्ताव करने को पेंच कहते हैं। पलटे में पेंचों के अक्षर भी बदलते हैं।

When we play a Baant, Kaayda or other forms of Bandish
Parans with the switching of words keeping the same structure
and repeat them in different Laya variations, it's known as a
Pench.

Rela

रेला

थोड़े से शब्दों की बार बार रुकते हुए कहने की चाल की क्रिया को रेला कहते हैं।
इसमें प्राय: शब्दों का परिवर्तन नहीं होता।

१	२	३	४	५

यथा :- धा तिरकिट तक धा तिरकिट तक धा तिरकित तक तूना किटतक ता

६	७	८

तिरकिट तक ता तिरकिट तक धा तिरकिट तक तूना किट तक ।

"Rela" literally means a stream or a fast flow of water. Rela
maintains a very important part in Taal Prastar, in both, Solo
as well as accompanying formats. The Solo concert almost
always ends with a fast Rela of few words with a continuous
flow than a brief pause and again continuous flow.

In accompanying formats such as Vocal Kirtans played in
temples of Brij Bhoomi in India, the main Pad is sung in
Vilambit or Madhyam Laya as per the mood of the Kirtan.
Then the last line which is usually called the "tuuk" where the
poet imprints his/her signature in the words, the "chalti" drut
portion begins where Rela is used very effectively with
Jhaanjh (cymbals).

149

Ladi (Dhanakshari Chand)

लड़ी

इसे घनाक्षरी छन्द भी कहते हैं। जिन शब्दों का रूप एक दूसरे से मिलते हुए लगातार चला जावे ऐसी क्रिया को लड़ी कहते हैं । यथा :-

१	२	३	४	५	६	७	८
तकिट	तकिट	धिकिट	धिकिट	तकिट	तकिट	तक	धिकिट	धिकिट	तक	धिकिट

९	१०	११	१२	१३	१४	१५ १६	१
तकिट तकिट धिकिट तक कत कत तक कत तक तक कत धाकत धाकत धा।

This is a very beautiful and simple yet effective technique. Ladi suggests small synonymous sounds and phrases of Bols played next to each other sequentially and alternate forms. This creates a very pleasing flow of Laya and in Mridang especially this is the same as Dhantakshari Chand. (Chand and their descriptions are given in Hindi in next advanced section)

Laggi

लग्गी

लग्गी अर्थात् (बाँस) नीचे मोटा होता है ऊपर पतला होता चला जाता है, इसी प्रकार जो भी शब्द एक से कुछ दूरी लेते हुए आगे दूसरे छोटे शब्दों को रखते हुए अर्थात् पहले की दूसरी से कम दूरी लेते हुए हों इस प्रकार के शब्दों के निर्माण की क्रिया को लग्गी कहते हैं।

दूसरा प्रकार - मध्य लय की गति को लेते हुए शब्दों की रचना को करते हुए द्रुत गति की चाल शब्दों को छोटा करते हुए क्रिया करने को लग्गी कहते हैं। उदा० -
लग्गी नं० (१)

१	२	३	४	५	६	७	८		९
दाग किट तक ना तिरकिट तिरकिट तीना तित् तिरकिट तक धिर किट तक

१०	११	१२ १३	१४	१५
कत् तिरकिट तक धिर किट तक धा ऽ तिरकिट तक धिर किट तक धा ऽ

१६		१
तिरकिटतक धिरकिटतक धा

लग्गी नं० (२) धागे ना धा तिरकिट धागे ना धा तिरकिट, धागेन धागेन धागे नागे तागेन धातीना तगेन तगेन धागे नागे धागेन धाधीना, धीक् धीना धिनगिन धागेन ताग् तीनाड़ा तीक् तीना तिन गिन धागेन धाग् धीनाड़ा, धीक् धीना धीक् धीना धीक् धीना नाधी धीना, तीक् तीना तीक् तीना तीक् तीना नाधी धीना,

Laggi – is derived from the word "Baans" which in English

refs to Bamboo. The Bamboo is swollen at bottom but then keeps growing in slender fashion. **Similarly here in Laggi what happens is that we start with Madhyam Laya phrases and then the Laya becomes Drut as the Laggi evolves and therefore the words contained in drut formations become shorter and hence the flow proceeds.**

Uthaan (Amad)

उठान अथवा आमद

तबले के शब्दों की ऐसी रचना जिसमें तीन लयों का अर्थात् बराबर दून और चौगुन का व्यवहार करके सम पर एक दम गाया जाता है। इसमें प्राय: तिहाई नहीं होती है। तबले में लहरे के साथ अथवा संगत करने के समय प्रारम्भ में आमद बजाई जाती है। कहीं-कहीं तिया लगाकर भी गाते हैं किन्तु तिया की भी गति का परिणाम भाजन इस परिभाषा के अनुसार ही होना चाहिए।

This is similar to the Mohra as it is played in the beginning but Uthaan is more powerful than a Mohra and has more structure. The Uthaan (meaning Getting Up or starting something) contains such a composition whereby Madhyam (Barabar) Laya is followed by Dugun Laya and then Chaugun Laya and then arrives at SAMA with final emphasis on Dha. Usually this does not contain a Tihai but sometimes if it does, then Tihai also inside it must have the same Triple Laya combination.

Section 4 - Advanced Chand Formations

In this section for the "Sadhak" and student of Indian Taal, we give some examples of Chands in Rhythm. A "Chand" in Hindi means a poetic meter of composing verse. In Taal Science, the Chands play a crucial role in creating a good accompaniment to the corresponding poetic Chands in the spoken vocal music. This advanced section is for a student who is already familiar with the Taal foundation and Hindi language, hence we assume this prerequisite knowledge.

We give these Chands so that they can be applied in practice to improve one's understanding of many different Laya and Yati (two of the ten Prana of Taal described earlier) combinations used in composing Chand meters in Rhythm. Translating these in the English language would be essentially counterproductive to our main aim in this book to discuss the foundational science and analysis of the entire Indian Rhythmic Taal theory. The advanced section is for those willing to put the science and theory in practice and continue their progress with the ideas described here.

Chand Bhujang Prayat

छन्द भुजंग प्रयात
यचौं मैं प्रभृते यही हाथ जोरी ।
फिरे आप-ते न कबौं बुद्धि मोरी ।।
भुजंग प्रयातो पमा चित्त जाको ।
जुरै ना कदा भूल कै संग ताको ।।

बोल तिताला ताल का छन्द भुजंग प्रयात

+ | o | +

१ २ ३ ४ ५ ६ ७ ८ ९ १० ११ १२ १३ १४ १५ १६ १

कृधान कतृत् धान ताकृ तान कतृत् धान धाकृ धान कतृत् धाकृ धान कतृत् धाकृ धान कतृत् धा

Jhulna Chand (Three varieties)

First Variety

झूलना, छन्द तीन प्रकार का

झूलना, गीतिका, गीतांगी, गीता और अन्य नाम बैताल भी कहते हैं। झूलना
प्रथम और गीता का एक रूप है, गीतिका और गीतांगी का एक रूप है। प्रथम
झूलना का सूक्ष्म रूप (SI)

स्थूल रूप :- मुनि राम गुनि, बान युत गल,
झूलन प्रथम मति मौन ।
हरि सम विभु, पावन परम,
जन हिय वसत, रति जान ।।

ताल तिताले में, प्रथम झूलना और गीता का बोल

+ | o   +

१ २ ३ ४ ५ ६ ७ ८ ९ १० ११ १२ १३ १४ १५ १६ १

केति कति टधा ऽकि टक तिट तागे नागे तधा ऽत धा तधा ऽतधा तधा ऽत धा

गीतिका और गीतांगी का सूक्ष्म रूप (IS)

स्थूल रूप - रत्न रवि कल धार कैलग, अन्त रचिये गीतिका ।
क्यों विसारे श्याम सुन्दर, यह धरी अन रीतिका ।।

ताल तिताले में गीतिका और गीतांगी का बोल

+ | o   +

१ २ ३ ४ ५ ६ ७ ८ ९ १० ११ १२ १३ १४ १५ १६ १

केकिट कतकिटतक ककिट धाक ऽधान धाता तगेग किटकिट ककिट धाक ऽधाक तिट धाक ऽधाक तिट धाक ऽधा

153

Second Variety Jhulna Chand

द्वितीय झूलना

सैंतिस यगंत यति, दोष दष दोष मुनि ।

जान रचिये द्वितीय, झूलना को ।।

आठ मात्रा के तिताला ताल में द्वितीय झूलना का बोल

+				o				+
१	२	३	४	५	६	७	८	९

कातिरकिटधा ऽनकिट तिटकतिटधान कतिट ताऽनधिट तिटधगे नधा तिरकिटधा

Third Variety Jhulna Chand

तृतीय झूलना

तीन दस झूलना अन्त मुनि झूलना दोष पद तीसरी भेद भायो ।

राम भजु बावरे राम भज बावरे राम के नाम को वेद गायो ।।।

तिताल ताल में तृतीय झूलना का बोल

+						|			o		|				+
१	२	३	४	५	६	७	८	९	१०	११	१२ १३		१४ १५		१६ १७

धान धातिट कतिट तगन नातिट तगेन कतिट धातिट धानित धान कतिट तान धिकिट धान तान धान धा

"Ashwa Gati" Chand

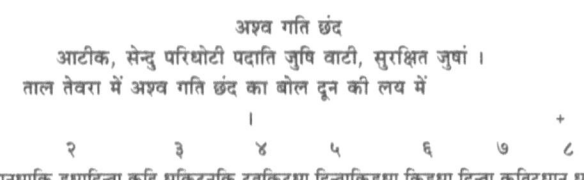

अश्व गति छंद

आटीक, सेन्दु परिधोटी पदाति जुषि वाटी, सुरक्षित जुषां ।

ताल तेवरा में अश्व गति छंद का बोल दून की लय में

+							+
१	२	३	४	५	६	७	८

धानधाकि ड्रधादिन्ता कदि धकिटनकि टतकिटधा दिन्ताकिड्रधा किड्रधा दिन्ता कतिटधान धा

Shikharini Chand

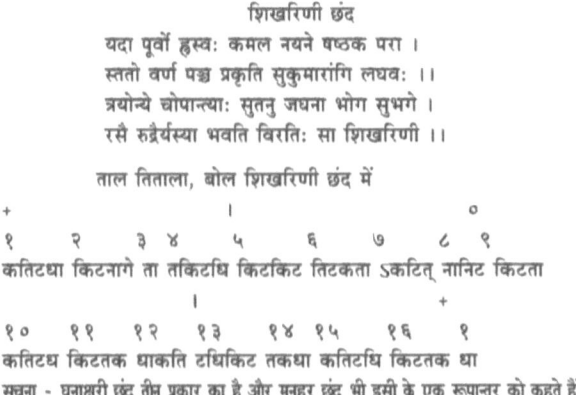

शिखरिणी छंद

यदा पूर्वो ह्रस्व: कमल नयने षष्ठक परा ।

स्ततो वर्ण पङ्ख प्रकृति सुकुमारांगि लघव: ।।

त्रयोन्ये चोपान्त्या: सुतनु जघना भोग सुभगे ।

रसे रुद्रैर्यस्या भवति विरति: सा शिखरिणी ।।

ताल तिताला, बोल शिखरिणी छंद में

+						o
१	२	३ ४	५	६	७	८ ९

कतिटधा किटनागे ता तकिटधि किटकिट तिटकता ऽकटित् नानिट किटता

		+
१० ११ १२ १३ १४ १५ १६	१

कतिटध किटतक धाकति टधिकिट तकधा कतिटधि किटतक धा

सूचना - घनाक्षरी छंद तीन प्रकार का है और मनहर छंद भी इसी के एक रूपान्तर को कहते हैं।

Manhar Chand

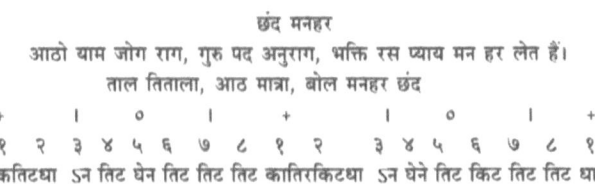

छंद मनहर

आठो याम जोग राग, गुरु पद अनुराग, भक्ति रस प्याय मन हर लेत हैं।

ताल तिताला, आठ मात्रा, बोल मनहर छंद

+		o		+		o		+
१	२ ३	४ ५	६ ७	८ १	२	३ ४	५ ६	७ ८ १

कतिटधा ऽन तिट घेन तिट तिट तिट कातिरकिटधा ऽन घेने तिट किट तिट तिट धा

Dhanakshari Chand

घनाक्षरी छंद
सुंदर सुजान पर मंद मुसकान पर, बाँसुरी की तान पर ठौरन ठगी रहे ।
मूरति विशाल पर, कंचन सी माल पर, हंसन सी चाल पर खोरन खगी रहे ।।

ताल तिताला आठ मात्रा का बोल घनाक्षरी छंद का

+			o		+		
१	२	३ ४	५ ६	७ ८	१	२	३

कातिर किटधा ऽन तिट तिट कृधा तिट धिट कृधा तिट धा

o		+
४ ५	६ ७	८ १

कृधा तिट धा कृधा तिट धा

रूप घनाक्षरी छंद
राम राम राम लोक नाम है अनूप रूप घन अक्षरी है भक्ति भव सिंधु हर जाल।

ताल तिताला ८ मात्रा का बोल रूप घनाक्षरी छंद

+			o		
१	२	३ ४	५ ६	७	८

कातिर किटधा तिट किड़धा तिट कत कातिर किटतक

+			o			+
१	२	३ ४	५	६	७	८ १

तागे तिट कत कत कातिर किटतक तागे तिट धा

Dev Dhanakshari Chand.

देव घनाक्षरी छंद
राम योग भक्ति मेव जानि जपें महा देव
घन अक्षरी सो उठे दामिनी दमकि दमकि
देव घनाक्षरी छंद में ८ मात्रा के तिताले का बोल

+			o			+
१	२	३	४	५	६	७	८ १

गेदिनता तिटकिट तागे नागे तिटकता का तिरकिटधा ऽनतिट किटगदि नागेदिन धा

Dandak Chand

छंद दण्डक

तेवरा, मत्तेभविक्रीड़ित और अन्य नाममुनि शेखर भी कहते हैं--
सुमरीना मयि लागती विलसति, मत्तेभविक्रीड़िता ।
मति ओछी जस धारती तस रहैं भारावहा पीड़िता ।।

छंद तेवरा वा दण्डिका का बोल ताल तिताला

+				o				+
१ २ ३ ४	५ ६ ७ ८	९ १० ११ १२	१३ १४ १५	१६	१

धागे न धाग दिन्ता कत कृ धानी धानी तागेन तागे दिन्ता कत कृ धानी धानी धा

Tevra Chand in Double Speed

दून की लय में छन्द तेवरा

+ |
१   २   ३   ४   ५   ६   ७   ८
धाकि टतक धुमतिर किटकधाकि टतक धुमतिरकिटतक

o | +
९ १०   ११   १२   १३ १४ १५ १६   १   २
ताकिटतक धुमतिर किटतक धाकिटतक धुमतिर किटतक धाकिट तकधुम

| o
३   ४   ५   ६   ७   ८   ९   १०   ११
किटतकि टतका किटतक धुमतिर किटतक धाकिट तकधुम किटतकि टतका

| +
१२   १३   १४ १५ १६   १
धुमतिर किटतक धाकिट तकधा किटतक धा

Gandaka Chand

गण्डका

ताल चामरध्वजं पयोधरं च कुण्डलं शरंविधाय ।
नूपुरं च नायकं सपक्षिराज गन्ध चामरंनिधाय ।।
रूपमन्त्यगं विधेहि वर्णिता च पन्नगेन्द्र पिंगलेन ।
गंडका कवीन्द्र मण्डली विनोदकारिणी सुमंगलेन ।।

बोल ठेका एक ताला

	+					+
१० ११ १२ १	२ ३ ४ ५६	७८ ९ १० ११ १२१

धान धातिट कतिट नातिट गेनेना नातिट कतिट ता कतिट तानता धाकति टतान ताधाक तिटता नता धा

Champakmala (Keherwa) Chand

चम्पकमाला वा कहरवा छंद कहते हैं।
तन्वि गुरु स्यादाद्य चतुर्थ पंचमषष्ठं चान्त्यमुपान्त्यम् ।
इन्द्रियवाणी यंत्र विराम: सा कमनीया चम्पकमाला ।।
बोल ठेका, एक ताला

```
  +                        |
  १    २    ३    ४    ५    ६    ७
कतितता ऽनकिट कातिरकिटथि किटधिट धगेना ऽनतिट नगेनता
```

```
       |                        +
  ८    ९    १०   ११   १२   १
ऽनधा दिन् दिन् तिटतिट धाकत धाकट धा
```

Charchari (Jhaptala) Chand

चर्चरी वा झपताला छंद

हारयुक्त सुवर्ण कुण्डलपाणि शङ्ख विराजिता ।
पाद नूपुरसर्गंता सुपयोधर द्वयभूषिता ।।
शोभिता वलयेन पन्नगराज पिंगल वर्णिता ।
चर्चरी तरुणी वचेतसि चाकसीति सुसंगता ।।
बोल ठेका एक ताला

```
  +                        |
  १    २    ३    ४    ५    ६    ७    ८
धागेतिट धागेतिट किटतागे तिटधागे तिटकिट धगेनध किटकिट ताधगेन
  |                        +
  ९    १०   ११   १२   १
धकिट किटता धगेनध किटकिट ता
```

वीणावादनतत्वज्ञ: श्रुतिजातिविशारद: ।
तालज्ञश्चाऽप्रयासेन मोक्षमार्गं प्रयच्छति ।।

Bibliography

Some of the ancient Musical texts and Philosophical texts referenced by Sw.
Shri Mannuji and present Authors in compilation of this volume are as follows;

1. Perspectives on Rasa ("Ras Drishti ki Tarfen me") by Shri Shyam Manohar
   Goswamyji

2. Taal Deepika (4 volumes) by Sw. Shri Mannuji

3. Mridang Ank by Sw. Shri Pagalbaba

4. Nada Rasa by NL Shri Mukundray Goswamy

5. Natya Shastra by Sage Bharat

6. Natya Shastra Tika Commentary by Abhinav Gupt

7. Sangeet Ratnakar by Sarangadev

8. Sangeet Makarand by Sage Narada

9. Shri Subodhiniji Ras Panchadhyayi with Tika from Purushottamji (by Jagganath
   MoonMoonji Chaturvedi, Chaukhambha Prakasahan Benaras)

10. "Brahma Sutra" by Veda Vyasa and its commentaries by Shri Shyamubava and Shri
    Hariray Mahaprabhu

11. Discourses on Shri Subodhiniji and Brahmavaad "Laghu Granth Tika Sangrah" by
    Shri Dixitji Maharaj (father of Shri Shyamubava)

12. Measuring the representational space of music with fMRI: a case study with Sting,
    Daniel J. Levitin & Scott T. Grafton Pages 548-557 | Received 18 Apr 2016,
    Accepted 19 Jul 2016, Published online: 12 Aug 2016

13. Music on the Mind: an introduction to this special issue of Neurocase, Indre V.
    Viskontas & Elizabeth Hellmuth Margulis, Pages 484-485 | Published online: 21 Dec
    2016

14. Ross, Jessica & Iversen, John & Balasubramaniam, Ramesh. (2016). Motor
    Simulation Theories of Musical Beat Perception.. Neurocase. 22.
    10.1080/13554794.2016.1242756.

About the Author

BHAVESH C. BHAGAT

(VIRGINIA, VARANASI, AND VADODARA)

HTTP://LINKEDIN.COM/IN/BHAVESHBHAGAT
Twitter: @bbhagat (Yogi Entrepreneur)
Facebook: @PakhawajMridang
Facebook: @UniversalPilgrim
Facebook: @YogiEntreprenur

Author Bhavesh C. Bhagat is a Father, Entrepreneur, Independent Board Member, Electronics Engineer, musician, and Martial Artist. Above all, he is a "Universal Pilgrim" always seeking new inner and outer destinations to explore. Gratefully observing and continuously learning from the many varied experiences of Life, Places and the "Journey" itself. He has extensively traveled to many remote parts of the world and explored all continents of the world as a successful Entrepreneur, Chairman, and CEO as a Qualified Technology Executive for the Boards of Directors.

Now in his second innings of life, he is on an *"Inner Pilgrimage"* dedicating it to exploring the innermost hidden destinations of "Swara" and "Laya" in the journey of music. This second phase Journey in its fullness is to explore the spiritual dimension of Sound and Devotion. He is a student and practitioner of the Shuddhadvaita Branch of Hindu Vedanta philosophy (Vallabh Vedantin Pushti Sampradaya). He was honored to represent this Hindu philosophical school of thought in the 2018 Parliament of World Religions held in Toronto, Canada.

In his professional life, Bhavesh Bhagat is an Independent Board Member & Chairman of several global organizations. He is an International Public Speaker, Author & Expert on subjects of Cyber Security, Emerging Technologies, Governance & Risk, Environmental and Health and Sustainability. He is a Serial Entrepreneur and Inventor of Governance as a Service® cloud-computing platform. For Board of Director positions, he is a Qualified Technology Executive with two decades of experience as a global Chief Executive, Board Member & Advisor to various Profit and Non-Profit Boards in diverse Industries. He now advises CEO's and Boards on Creative and Agile form of Leadership leveraging diverse human experiences of music and Martial Arts. Bhavesh mentors other Startups and advises them on how to create good Cyber Security based businesses as a Board Level Mentor at Mach37 Incubator in Virginia.

He is a lifelong student and practitioner of Applied Philosophy in the form of Indian Classical music practicing oldest Indian Percussion Instrument Mridang (Pakhawaj). As a student of Vocal singing, he is focused on the practice of the Brij Bhasha Padas sung in Dhrupad and Khayal styles. He is also a student of the Indian Bamboo Flute (Bansuri). In this way, Bhavesh is striving in the present second innings of life to initiate the musical journeys on all three fronts, Taal, Swara, and Instrument. Each Taal and each Swara would take at least one lifetime to understand and that too in their basic forms, so Author has plenty of time and is in no rush to complete this musical Journey in this life.

Music for this Author is not an art, instrument or means to get "somewhere". Music for this Author is the irreplaceable joyous ingredient of offerings to the Lord.

Bhavesh is a 2nd-degree Kukkiwon Board-certified Black Belt Martial Artist in TaeKwonDo (South Korean) form of Martial Arts. The author has synthesized all his unique expertise in leading Global Organizations, music and Martial Arts to deliver and teach the Boards and CEO's the need to diversify the Leadership skills to continuously survive in face of social, technological and geopolitical winds of change we face with Agility and Creativity.

Other books by Author

Science of Melody – This book is an English synopsis with some of the Author's experiences and interpretations of the Science of Melody. The book deals with the subject of illustrating the Science (logical and experimental roots) and Sensibilities (emotional and spiritual feeling based genesis) of the Indian Melodic structures also known as "Swara Shastra" in Sanskrit. The book is a first of its kind in the English language to explore the synthesis of Science, Spirituality, and Art in the context of roots of Melody in Music and their structured systems of Indian Classical Musical. The knowledge is based on ancient texts and the author's evolution as a practitioner of the art and instructions received from and documented by profound Gurus over the years. The first analytical and scientific section is based on the Author's continuously evolving experiments in the practice of Naad Yoga and understanding the physical, spiritual and psychosomatic roots of Melody in Music. Second, third and fourth sections go into the ancient details of the structure and science of the Swara system in India with the English descriptions and explanations of the Sanskrit terms and their meanings. Source of the Sanskrit and English translated material is based on the ancient treatise Sangeet Ratnakar of Sharangdev from and its translation by Dr. R.K. Shringy in 1978.

The original Sanskrit verses have also been maintained in this book to assist the reader in grasping the idea from multiple languages. The purpose of the book is to act as a reference and inspiration to educate the practitioner of Music and Melody on true science and sensibilities of the Indian Melody structures so that one can perform with the full depth that is gained by exploring the subject from all avenues. This is a second volume in the Indian Naad Yoga series first of which deals with Science of Rhythm and the reader is advised to refer to both volumes for the complete understanding of the Indian Musical system.

Also Available on Amazon Kindle Unlimited – Science of Melody Kindle Edition

About the Publisher

Universal Pilgrim Productions (UPP) is a global creative and media-independent publisher with a mission for unifying the scientific, artistic, philosophical and spiritual dimensions of our human existence. Having Eastern and Western readers in mind, at UPP, we believe in an infinitely joyous journey of seeking knowledge. A **"universal pilgrimage"** where the physical destination is irrelevant.

We commit our publishing projects to a Pilgrimage of human existence... of any eager soul striving to make efforts on the path of learning. We strive to publish multiple series of rare out of print and important global cultural assets in the present digital medium of Kindle and other future media as they evolve. Our mission is to benefit humanity by sharing of global cultural values and literary assets. *For future projects that fit our mission, please do contact us via email pilgrimuniverse@gmail.com*

We strive to use the latest platforms such as Amazon and other present and future media to publish at the lowest cost allowed by the platform with a predominantly non-profit mission. Where applicable all our publications depending on their nature would be disseminated free of cost using the Kindle Unlimited global platform for Kindle readers. *For all non-profit projects, we commit to our love for animals that any nominal material proceeds would be donated to the betterment of Cows in North America.*

The present volume is the new series of publications under the "Nada Yoga" series by UPP and is disseminated in the non-profit model with all proceeds donated towards Cow welfare in North America. All future volumes in this "Naad Yoga" series will have the same non-profit mission.

UPP has published several volumes on Amazon in the ongoing non-profit "Vallabh Vedant" series focusing on the most important and significant Eastern Hindu Philosophical works in Hindi, Brij Bhasha, Gujarati and English languages of Mahaprabhu Shri Vallabhacharya and his Pure Non Dualistic philosophy "PushtiMarg - Path of Divine Grace." Our humble devoted effort is to forever make these Kindle centered publications available at no cost.

Virginia, Varanasi, Vadodara

May the Light Guide the "Laya" of the Seeker
Tulsi Ghat, Varanasi, Photo by Author

A miniature from Author's personal collection depicting Kangra style of Mridang and Dance movements of Lord Shri Radhe Krishna acquired in Rajasthan, India in 2007.

www.ingramcontent.com/pod-product-compliance
Lightning Source LLC
Chambersburg PA
CBHW021543200526
45163CB00015B/846